Chemistry of
Diazirines

Volume II

Editor

Michael T. H. Liu, B.Sc., Ph.D.

Professor
Department of Chemistry
University of Prince Edward Island
Charlottetown, P.E.I., Canada

CRC Press, Inc.
Boca Raton, Florida

Library of Congress Cataloging-in-Publication Data

Chemistry of diazirines.

Includes index and bibliographies.
1. Diazirines. I. Liu, Michael T. H.
QD341.A9C54 1987   547'.043   87-10276
ISBN 0-8493-5047-6 (set)
ISBN 0-8493-5048-4 (v. 1)
ISBN 0-8493-5049-2 (v. 2)

Direct all inquiries to CRC Press, Inc., 2000 Corporate Blvd., N.W., Boca Raton, Florida, 33431.

©1987 by CRC Press, Inc.

International Standard Book Number 0-8493-5047-6 (set)
International Standard Book Number 0-8493-5048-4 (v. 1)
International Standard Book Number 0-8493-5049-2 (v. 2)

Library of Congress Card Number 87-10276
Printed in the United States

PREFACE

The chemistry of diazirines has a relatively short history, dating from the discovery of these compounds in 1960. Since then, the cyclic diazo compounds with three-membered rings (diazirines) have been added to the ranks of the well-known aliphatic diazo compounds. The past 25 years have witnessed a high level of activity in the study of the chemistry of diazirines. As a result of many extensive investigations, diazirines, as a class, have become better understood.

Since the chemistry of diazirines has been the subject of increasing interest from several branches of chemistry, the time appears to be ripe to produce a monograph presenting the most recent innovations and discoveries in all aspects of diazirine chemistry. Although several review articles have been written on this subject, it is still difficult to locate the required information on the synthesis of diazirines or on general topics. The 11 chapters cover all areas of diazirine chemistry, including theoretical treatments, spectroscopy, synthesis, exchange reactions, thermolysis, photolysis, carbene selectivity, energetics, transition metal complexes, and electrochemical aspects.

Baird opens the first volume with a discussion on the theoretical aspects of diazirines. In Chapter 2, Winnewisser, Möller, and Gambi discuss the UV and IR spectra of diazirines. In Chapter 3, Schmitz describes the various synthetic methods for diazirines as well as some of their reactions. The exchange reactions of halodiazirines are presented by Moss in Chapter 4. Liu and Stevens deal with thermolysis and photolysis in Chapter 5, which is followed by diazirine-diazoalkane interconversions by Meier in Chapter 6. The dissociation energetics of simple diazirines is given by Okabe in Chapter 7. The selectivity of carbenes with particular emphasis on cyclopropanation reactions is discussed by Doyle in Chapter 8. Bayley, in Chapter 9, shows us some of the properties required to create efficient photoactivatable reagents in biochemistry. The transition metal carbonyl complexes from diazirines is presented by Kisch in Chapter 10. Finally, Elson and Liu discuss the electrochemistry of diazirines in the concluding chapter.

These volumes also present complete literature surveys and incisive commentary. The editor hopes that this work will be useful to chemists in a wide range of specialities.

I wish to thank several colleagues for their assistance and advice on various parts of the work. Dr. T. Spira has read the manuscript and provided many invaluable comments. Professor K. J. Laidler first introduced me to the study of chemical kinetics, and Professor H. M. Frey provided my first exposure to diazirine chemistry.

In closing, I wish to cite Sir R. G. W. Norrish's tribute to international scholarship, which accurately reflects the view of the editor of the present volume:

> "Science is international; none of the work described here would have happened but for work of others in countries 'round the world, for nature reveals herself to those with eyes to see and minds to comprehend, and such are not limited by geographical or political frontiers. Only by working together in pursuit of knowledge can we hope to achieve that mutual understanding upon which the well-being of our world depends."[1]

Michael T. H. Liu
September 1986

1. **Norrish, R. G. W.,** *Ann. Rev. Phys. Chem.,* 20, 1, 1969.

THE EDITOR

Michael T. H. Liu, Professor of Chemistry at the University of Prince Edward Island, obtained his B.Sc. degree from St. Dunstan's University, Charlottetown in 1961. Upon graduation with his M.A. in Chemistry (first class) from St. Francis Xavier University in 1964, he was group leader in R & D at Chemcell Ltd., Quebec, but later returned to his studies with Professor K. J. Laidler and was awarded his Ph.D. in 1967 with great distinction from the University of Ottawa.

In the same year he received a National Research Council Overseas postdoctoral fellowship tenable at the University of Reading. He joined the Chemistry Department at the University of Prince Edward Island the following year.

Dr. Liu was visiting professor at the University of British Columbia and the University of Geneva in 1975 and 1982, respectively. He was adjunct professor at Dalhousie University from 1979 to 1985. In 1980, he was named Fellow of the Chemical Institute of Canada. He is a member of the latter Institute and of the Inter-American Photochemical Society.

Dr. Liu has published over 70 articles and presented more than 20 papers at scientific meetings. He has lectured at universities and research centers around the globe.

Dr. Liu's current research interests include the synthesis of new diazirines, 1,2 hydrogen migration and oxidative studies of carbenes, cyclopropanation of electrophilic carbenes with electron-deficient olefins, and low temperatue photolysis of diazirines.

CONTRIBUTORS

Hagan Bayley, Ph.D.
Assistant Professor
Howard Hughes Medical Institute
College of Physicians and Surgeons
Columbia University
New York, New York

Horst Kisch
Professor
Institut für Anorganische Chemie
Universität Erlangen-Nürnberg
Erlangen, West Germany

Michael P. Doyle, Ph.D.
Dr. D. R. Semmes Distinguished
 Professor
Department of Chemistry
Trinity University
San Antonio, Texas

Michael T. H. Liu, B.Sc., Ph.D.
Professor
Department of Chemistry
University of Prince Edward Island
Charlottetown, P.E.I., Canada

Clive M. Elson, B.Sc., Ph.D.
Chairman
Department of Chemistry
Saint Mary's University
Halifax, Nova Scotia, Canada

Herbert Meier, Dr. rer. nat.
Professor of Chemistry
Institute of Organic Chemistry
University of Mainz
Mainz, West Germany

Hideo Okabe, Ph.D.
Visiting Professor
Department of Chemistry
Howard University
Washington, D.C.
Chemical Kinetics Division
National Bureau of Standards
Gaithersburg, Maryland

DEDICATION

This book is dedicated to the thousands of scientists whose names appear in the reference section of each chapter and to my wife Betty Y. L. and our children David C. Y., Stephen C. M., and Peter K. Y.

TABLE OF CONTENTS

Volume I

Volume II

Chapter 6

DIAZIRINE-DIAZOALKANE INTERCONVERSIONS

Herbert Meier

TABLE OF CONTENTS

I. INTRODUCTION

The reversible valence isomerization between diazo compounds **1** and 3H-diazirines **2** shows a new dimension of the historical dispute concerning the open-chained[1-3] or cyclic[4,5] structure of aliphatic diazo compounds. The number of direct interconversions **1** $\rightleftharpoons$ **2** is restricted by the fact that both isomers easily lose nitrogen by thermal or photochemical activation. One must also recognize the process of indirect isomerization by an elimination-addition process. This is especially applicable to the route **2** $\rightarrow$ **1**. It is well-known that under certain conditions carbenes **3** can add to nitrogen to form diazo compounds.[6] A survey of the currently known transformations is given in Scheme 1.

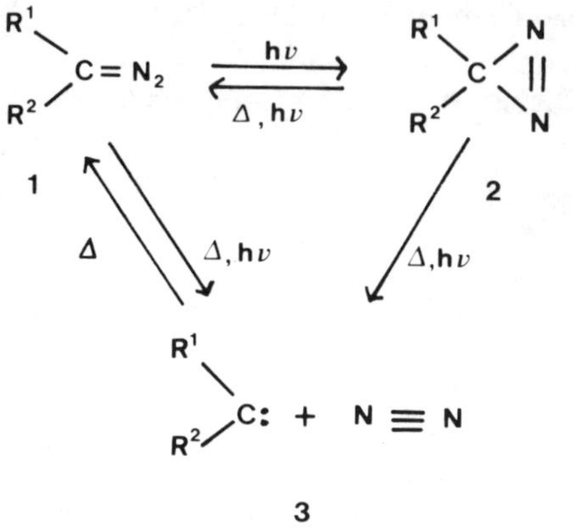

SCHEME 1.

In the case of α-diazocarbonyl compounds **4** and 3-acyl-3H-diazirines **5** the situation is further complicated by the possible contribution of further valence isomers.

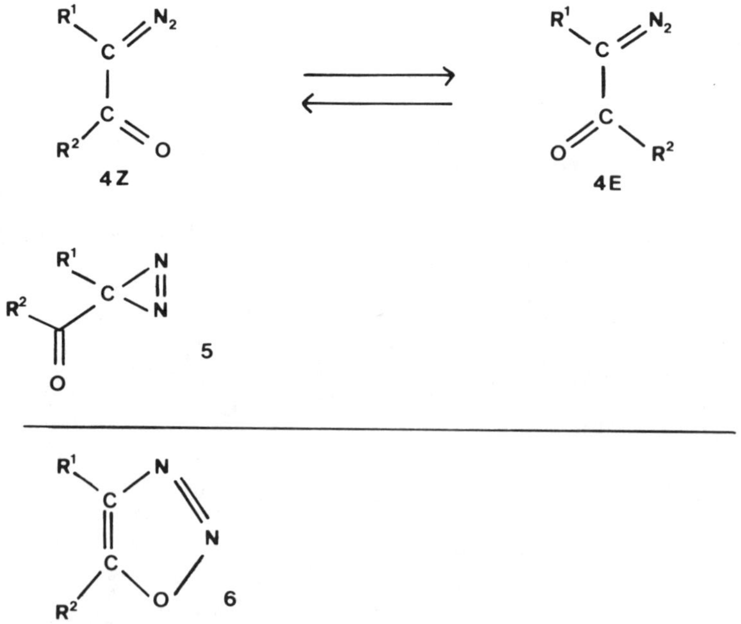

SCHEME 2.

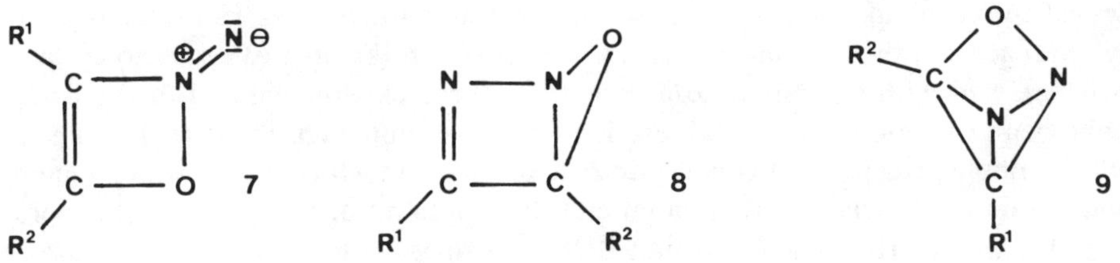

SCHEME 2 (continued)

Whereas there are no hints for the energetically high lying diazene structure **7** and the strained bicyclic and tricyclic species **8** and **9**, a 1,2,3-oxadiazole **6** could be established. 1,2,3-Benzoxadiazole **10** predominates in the equilibrium with 6-diazo-2,4-cyclohexadien-1-one **11** in the gas phase[7] or in nonpolar solvents.[8] **10**, isolated in an argon matrix at 15 K, is transformed by irradiation (λ = 300 nm) to **11**, which eliminates nitrogen upon further irradiation ($\lambda \geq$ 350 nm).[8]

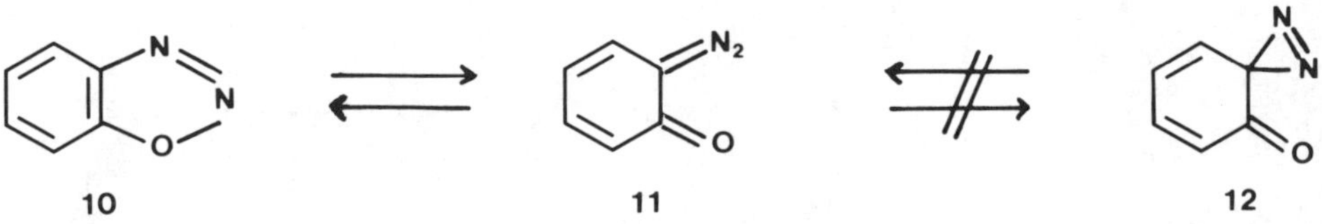

SCHEME 3.

II. CYCLIZATION OF DIAZO COMPOUNDS TO DIAZIRINES

In 1971, Lowe and Parker[9] first transformed a diazo compound into a diazirine. Diazoacetamide and some of its derivatives, with one or two substituents on the amide nitrogen atom, upon irradiation with visible light furnish the corresponding 3-aminocarbonyl-3H-diazirines.[9,10]

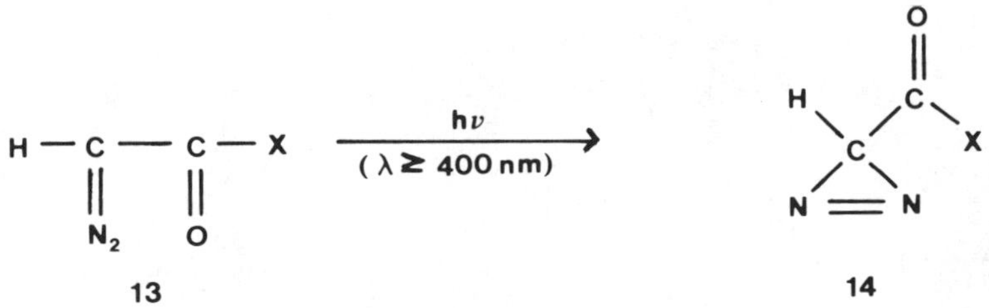

13/14	a	X =	NH_2	Low yield
	b		$NH(CH_3)$	31%
	c		$N(C_2H_5)_2$	22%
	d		$N(CH_3)\text{-}C_6H_5$	21%
	e		$N(C_6H_5)_2$	16%
	f		Piperidyl	20%

SCHEME 4.

The reaction, which can also be performed with N-diazoacetyl-L-proline benzyl ester and N-diazoacetyl-L-phenylalanine methyl ester, fails with N-(ethoxycarbonyldiazoacetyl)pyrrolidine, N-(*t*-butoxycarbonyldiazoacetyl)-L-proline benzyl ester, and N-(*t*-butoxycarbonyldiazoacetyl)piperidine.[9,10] Negative results were also obtained with some α-diazoesters (ethyl diazoacetate, diethyl diazomalonate), some α-diazo ketones (diazoacetone, diazoacetophenone), and with diphenyldiazomethane.[9,10] These experiments led to the conclusion that nitrogen extrusion is the normal reaction for diazo compounds in the first excited singlet state S_1, with the exception of some α-diazoacetamides, in which a canonical structure with an interaction of the diazo carbon atom and the amide nitrogen atom would suppress the CN bond cleavage. Thus, monosubstituted 3H-diazirines are formed. Higher α-diazoamides should lead to the corresponding 3,3-disubstituted diazirines. The first examples of this type were found in the indole series.[11,12]

Excitation to a higher singlet state S_2 leads in every case to a nitrogen evolution. The quantum yields Φ for the denitrogenation of diazoacetophenone, e.g., are 0.6, 0.5, and 0.07 for the irradiation at λ = 250, 294, and 325 nm.[13] The S_1 state of diazoacetophenone is predominantly radiationless deactivated by an internal conversion to the ground state S_0; a substantial intersystem crossing or a photoisomerization does not occur. The dependency of the diazirine formation on the wavelengths of the radiation is demonstrated by the following examples.[10,14,15]

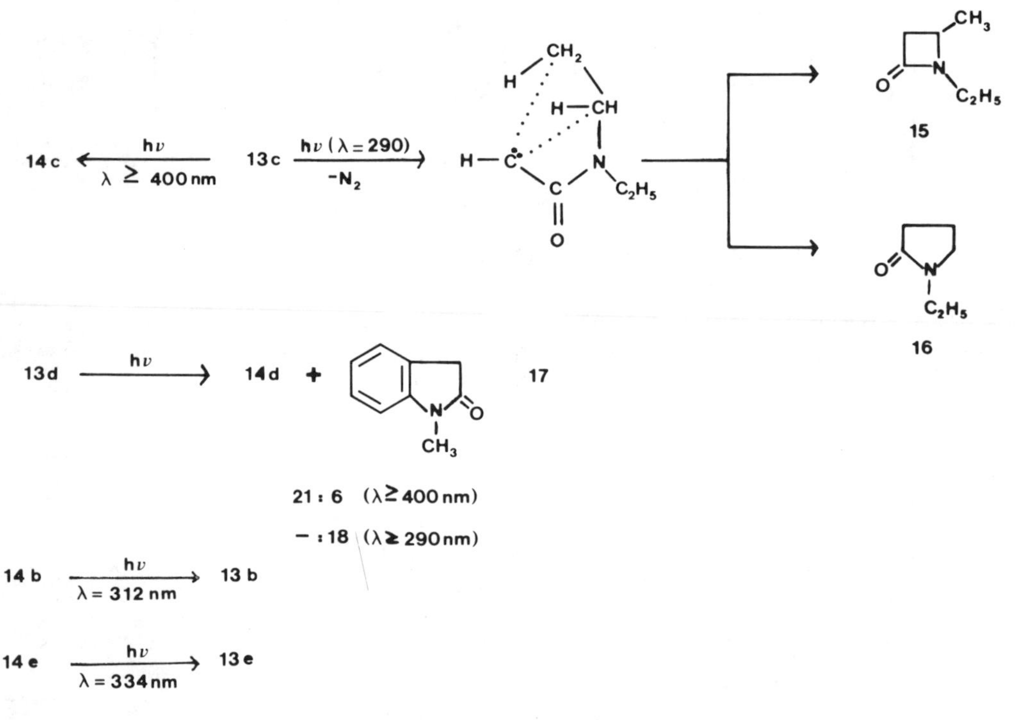

SCHEME 5.

14b and **14e** yield a photochemical back reaction on irradiation at λ = 312 and 334 nm. **13b** and **13e** are produced in a yield of 30 and 15%, respectively. The absorption range used corresponds to the $n\pi^*$ transition of the diazirine near 310 nm. However, it is interesting to note that **13e** is not formed on direct irradiation into the maximum probably due to an end absorption of another electron transition which already overlaps at 310 nm.

Preparative photolyses are often performed with cutoff filters. A Pyrex filter, for example, practically eliminates the radiation below λ = 290 nm. Under these conditions the S_1 and S_2 state of a diazo compound can be achieved. Hence, photoisomerization may occur from the S_1 state, directly populated by absorption $S_0 \rightarrow S_1$ or populated by $S_0 \rightarrow S_2$ and a fast internal conversion $S_2 \rightarrow S_1$. However, the diazirine formation may be overlooked in such an experiment because of the high efficiency of the nitrogen extrusion induced in the S_2 state, or because of the labile behavior of the diazirine. An illustration is given by the valence isomerism of 3-diazo-2-oxoindoline **18** and the corresponding 2′-oxospiro[diazirine-3,3′-indoline] **19**.[12]

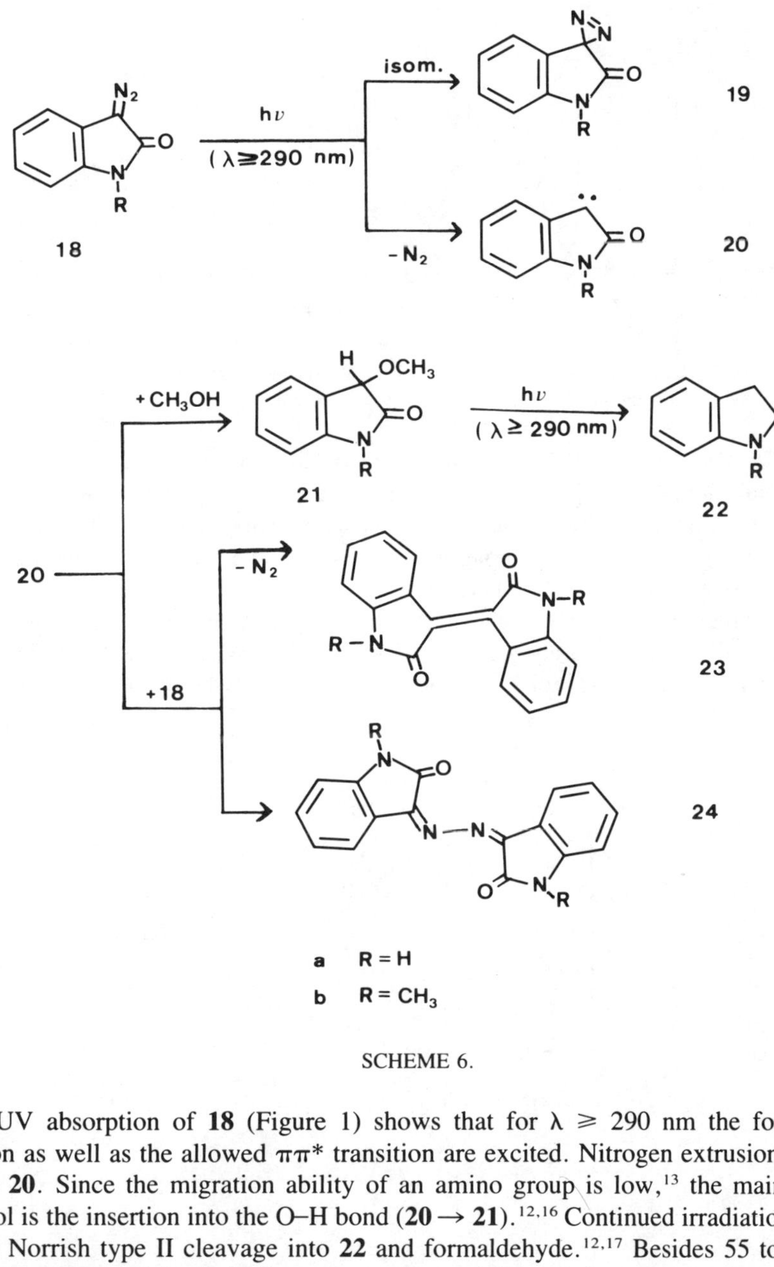

a   R = H

b   R = CH$_3$

SCHEME 6.

The UV absorption of **18** (Figure 1) shows that for $\lambda \geq 290$ nm the forbidden $n\pi^*$ transition as well as the allowed $\pi\pi^*$ transition are excited. Nitrogen extrusion leads to the carbene **20**. Since the migration ability of an amino group is low,[13] the main reaction in methanol is the insertion into the O–H bond (**20** → **21**).[12,16] Continued irradiation transforms **21** by a Norrish type II cleavage into **22** and formaldehyde.[12,17] Besides 55 to 70% of **21/22** small amounts of the isoindigo system **23** ($\leq 1\%$) and the azine **24** (3 to 4%) can be isolated.

Another reaction path, commencing with the first excited singlet state S_1, leads to the diazirine **19**. A plot of the concentrations of **18**, **19**, and **21** (R=CH$_3$) vs. the reaction time indicates that **19** is enriched during the room temperature photolysis to ~15 mol% (Figure

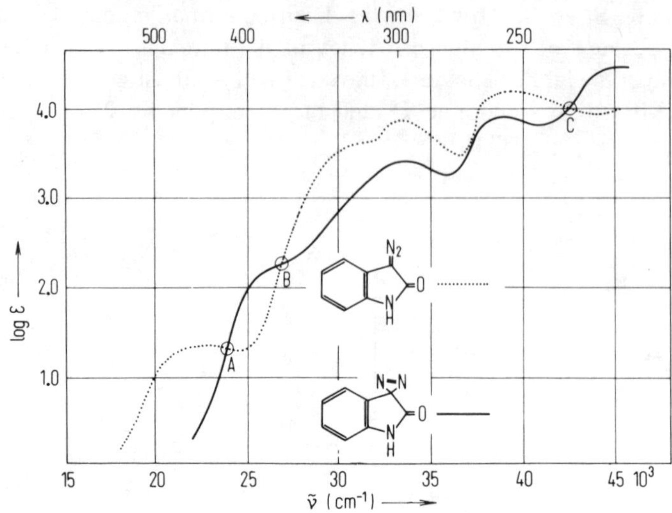

FIGURE 1.   UV/VIS spectra of 3-diazo-2-oxoindoline **18** (R=H) and 2′-oxospiro[diazirine-3,3′-indoline] **19** (R=H) in methanol.[12]

2). Eventually, the diazirine **19** vanishes in favor of **21**. The reverse reaction **19** → **18** is due to a thermal and a superimposed photochemical process. The photochemical ring opening is induced by the nπ* excitation of the diazirine near 380 nm (Figure 1). The photochemical reaction does not play a significant role at room temperature because the thermal route is much more efficient. Details of the thermal process are discussed elsewhere.

The diazo compound **25** shows a photoreactivity which resembles the diazoamides.[18] As a consequence of the low migration ability of alkoxy groups[13] cycloaddition processes of the intermediate carbene and not the Wolff rearrangement, compete with the diazirine formation.

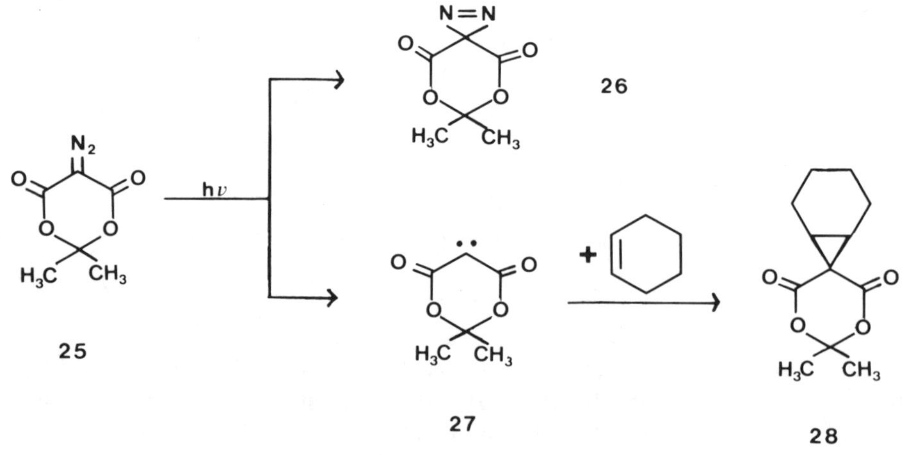

λ = 350 nm :   35 % **26**, 9 %   **28**
254 nm :   4% **26**, 84% **28**

SCHEME 7.

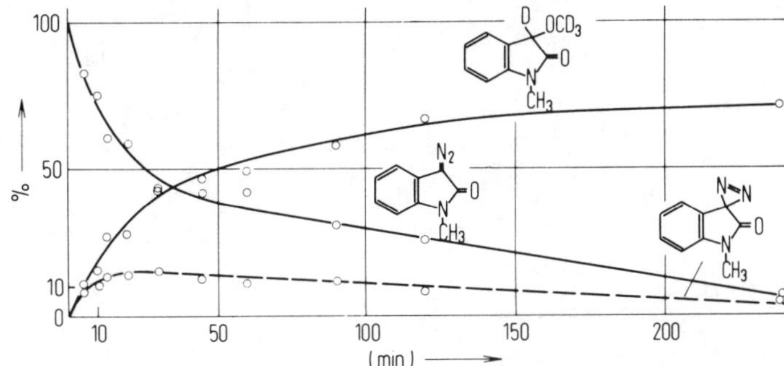

FIGURE 2.   Photolysis of 3-diazo-1-methyl-2-oxoindoline **18** (R=CH$_3$) in CD$_3$OD (27°C, λ ≥ 290 nm, mol% without calculating the solvent).

However, the photochemical isomerization reaction of diazo compounds to diazirines is not restricted to diazoamides or diazoesters. Irradiation of the α-diazoketones **29** and **31** furnishes 10 to 16% of the corresponding diazirines **30** and **32**.[19]

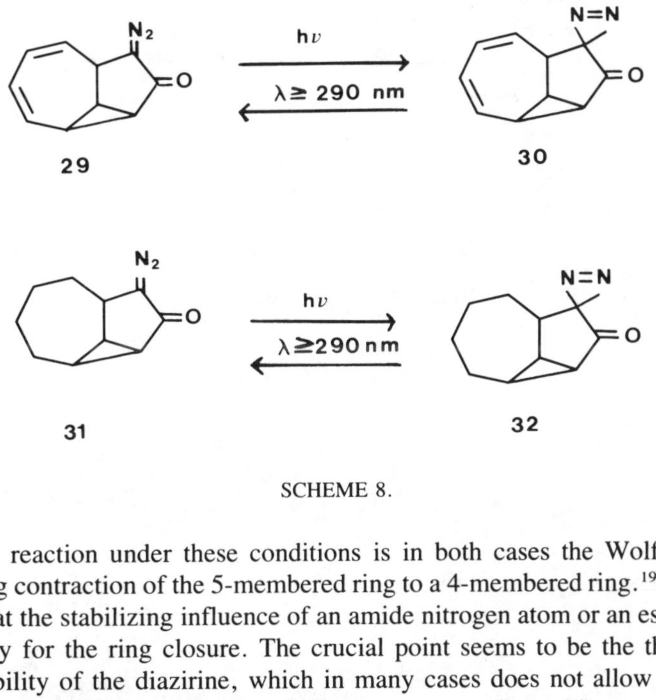

SCHEME 8.

The primary reaction under these conditions is in both cases the Wolff rearrangement leading to a ring contraction of the 5-membered ring to a 4-membered ring.[19] These examples demonstrate that the stabilizing influence of an amide nitrogen atom or an ester oxygen atom is not necessary for the ring closure. The crucial point seems to be the thermal and photochemical stability of the diazirine, which in many cases does not allow the detection of the diazirine isomer. Compounds **30** and **32** are unexpectedly stable compared with other 3-acyl-3H-diazirines such as 1-oxospiro[cyclohexane-2,3′-diazirine].[20]

The stabilizing effect of trifluoromethyl groups on small ring systems is well known. Irradiation of the diazoketone **33** at λ ≥ 320 nm in solution at room temperature or in an argon matrix at 10 K results in the formation of the fairly stable diazirine **34**. According to the ^{19}F-NMR spectrum the yield amounts to 40 to 45%. Besides Wolff rearrangement to *bis*[trifluormethyl]ketene **35** the participation of the solvent in addition reactions can be observed. Long wavelength irradiation in the gas phase gives, among others, **35**, and its isomer perfluormethacrylyl fluoride **36**, but no diazirine **34**.[21]

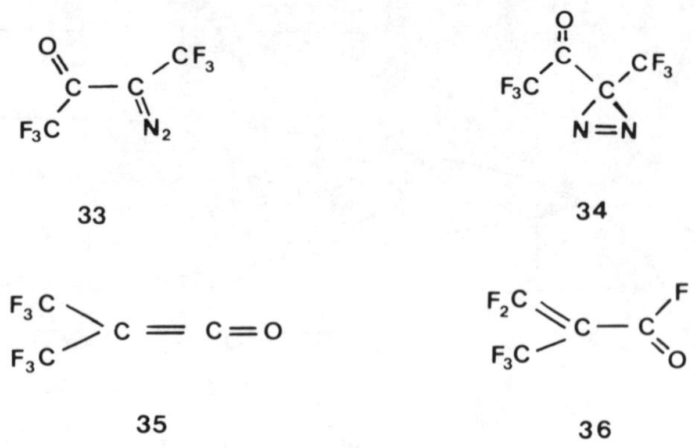

SCHEME 9.

There also exist some examples of diazo compounds without an α-carbonyl group, which show photochemical transformations to the corresponding diazirines. A photostationary state was found for the irradiation of trimethylsilyldiazomethane **37**/trimethylsilyldiazirine **38** in a matrix photolysis at 8 K and λ ⩾ 355 nm.[22,23]

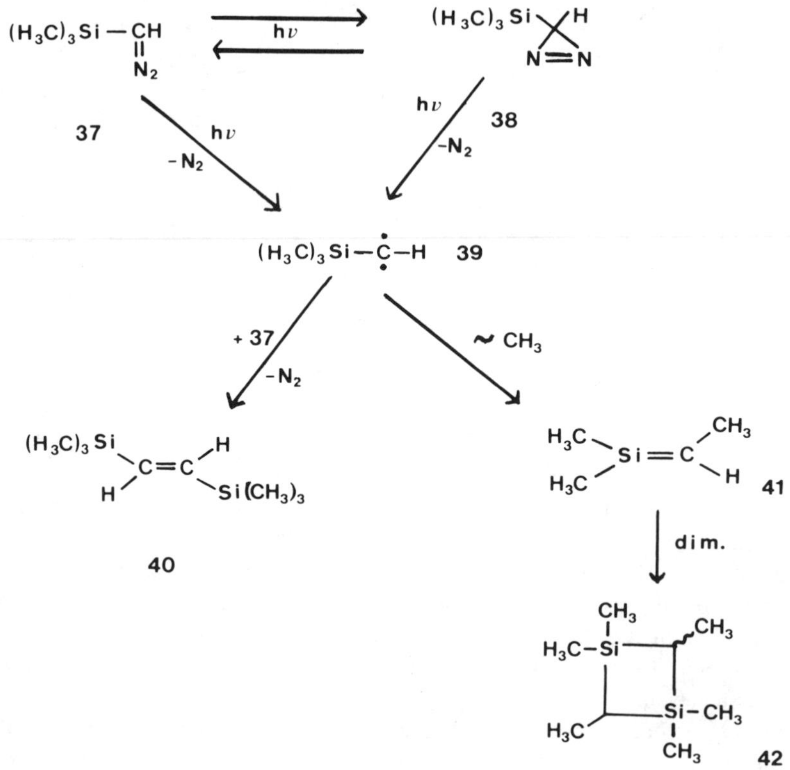

SCHEME 10.

Trimethylsilyldiazirine **38** is also formed upon irradiation in a variety of solvents at room temperature. The competitive elimination of nitrogen at shorter wavelengths leads to the carbene **39**, a ground state triplet. **39** is thermally stable between 4 and 40 K. Warming the matrix slowly to room temperature furnishes the olefin **40**. The singlet carbene, primarily formed in the continued irradiation of the photostationary mixture, rearranges to the silabutene **41**. The dimerization to *cis-* and *trans-*1,1,2,3,3,4-hexamethyl-1,3-disilacyclobutane (**41** → **42**) is possible, when the argon matrix loses its rigidity by warming to 40 or 45 K.

Investigations of the silabenzene and the 5-silafulvene system are closely related to the rearrangement **39** → **41**. Diazo compound **43** forms 64% of the diazirine **44** ($\lambda \geq 350$ nm, filter solution of phenanthrene in methanol). Both isomers eliminate nitrogen by irradiation at shorter wavelengths. The carbene **45** rearranges via a ring enlargement to the silabenzene **46** or by a methyl shift to the 5-silafulvene **47**. In alcoholic solution addition products of **46** and **47** can be isolated.[24]

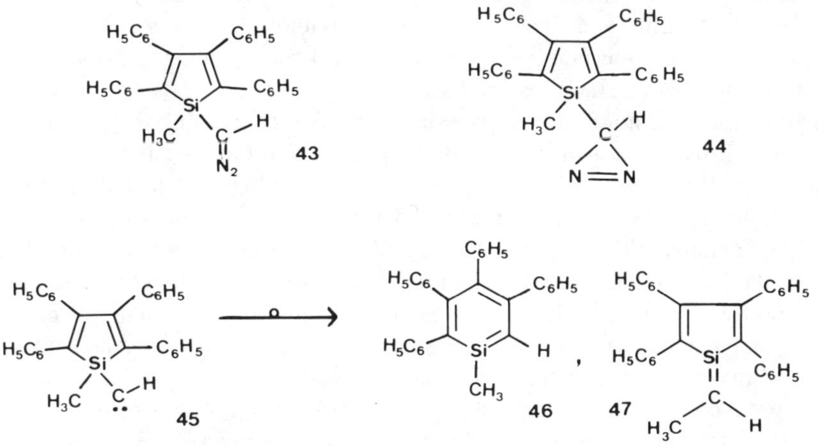

SCHEME 11.

Further examples of wavelength-dependent interconversions have been found in the series of fluorinated diazoalkanes.[25] The diazo compound **48** is transformed at $\lambda = 410$ nm near the maximum of the $S_0 \to S_1$ absorption, situated in cyclohexane at $\lambda_{max} = 390$ nm ($\epsilon_{max} \approx 12$) to the diazirine **49**. The yield of approximately 50% is confined by the competitive photofragmentation. The carbene **50** shows a 1,2 shift of the pentyl group (**50** → **51**) and an intramolecular insertion into the C–H bonds α to the difluoromethylene groups (**50** → *cis*-**52** and *trans*-**52**).

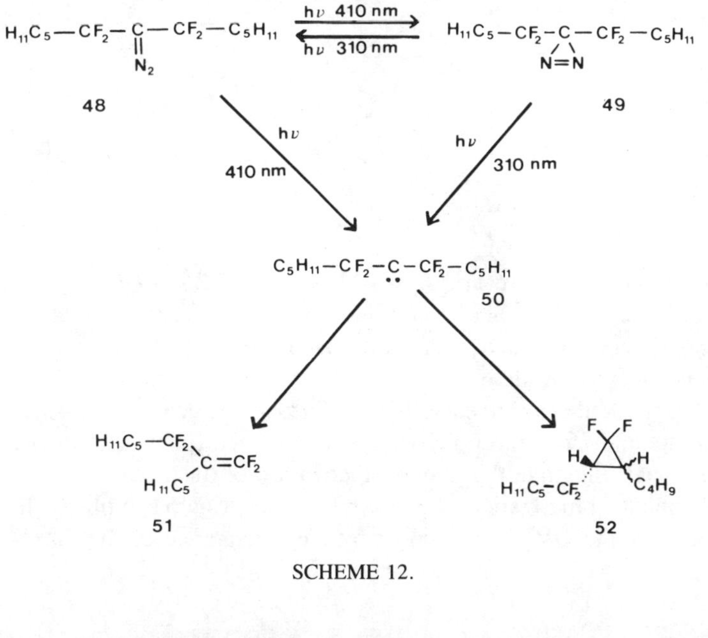

SCHEME 12.

The photolysis of **49** at 310 nm (λ_{max} in cyclohexane: 302, 312 nm, ϵ_{max} = 53) yields 56% **48**, 13% **51**, 6% *cis*-**52**, and 21% *trans*-**52**. Irradiation at 260 nm leads to the complete disappearance of **48** and **49**. Similar results have been obtained for 3-(1,1-difluorooctyl)-3H-diazirine and 3,3-*bis*[chlorodifluoromethyl]diazirine.[25]

III. RING OPENING OF DIAZIRINES TO DIAZO COMPOUNDS

In the preceding section numerous examples were discussed in which photochemically closed diazirine ring systems could be reopened by changing the wavelength of irradiation. Under suitable conditions, where there is an overlap in the absorption bands for the $S_0 \rightarrow S_1$ transitions of diazirine and diazo compound, a photostationary state may be reached. However, on the whole, the number of photochemical ring openings of diazirines is much higher than the number of ring closure reactions.[26]

The early literature featured a controversial discussion concerning the photochemistry of the unsubstituted diazirine. Amrich and Bell[27] photolyzed diazirine in the gas phase (5 to 30 torr diazirine, 0 to 600 torr N_2) using light of 320 nm. They detected diazomethane in the UV spectrum, but its formation accounted for only 10% of the quantum yield of the diazirine disappearance ($\Phi \approx 2$); i.e., at least 20% of the primary decomposition of diazirine proceeds on the isomerization route. The remaining part decays directly to methylene and nitrogen. Moore and Pimentel[6] photolyzed diazirine in a matrix enriched in ^{15}N and concluded that the diazomethane is formed in a secondary reaction of methylene and nitrogen. The generation of diazomethane in the gas phase photolysis can be totally quenched by the addition of propane (15 torr of propane per 3 torr of diazirine). This result was also taken as an indication that diazomethane cannot be a primary product.[28] Frey[26] pointed out that some energetical problems are also involved because directly formed diazomethane should exhibit an extremely high vibrational excitation, therefore indicating a fast decay rate. Nevertheless, the formation of diazoalkanes **1** as intermediates in the photochemical cleavage of diazirines **2** to carbenes **3** and nitrogen is basically accepted. In most diazirines, the reaction sequence $2 \rightarrow 3 \rightarrow 1$ seems to play a minor role.[6] Furthermore, it is interesting to note that a "side-on" attack of a carbene on the nitrogen molecule leading to a diazirine has not been detected.

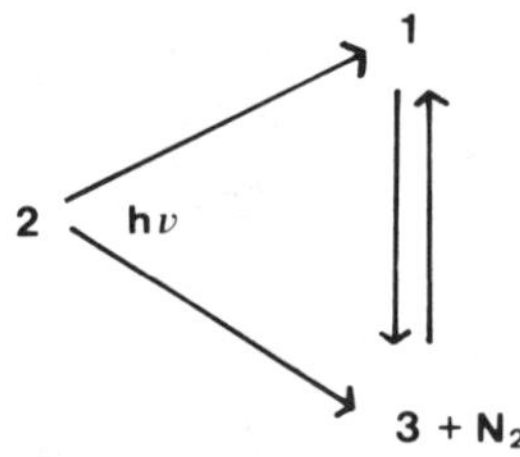

SCHEME 13.

The ratio of the concurrent reaction paths $2 \rightarrow 1 \rightarrow 3/2 \rightarrow 3$ strongly depends on the particular system and on the reaction conditions. This statement should certainly include borderline cases where the diazoalkane route is the only route, or on the other hand a diazoalkane is not formed at all.

There are several possibilities to establish the diazoalkane generation: isolation or trapping of the diazo compound, spectroscopic or kinetic evidence for its intermediate occurrence. A few examples shall illustrate the photoisomerization of diazirines.

The photochemical isomerization of 3-phenyl-3H-diazirine **53** to phenyldiazomethane **55** can be monitored by the UV absorption or by the appearance of the diazo-NN stretching vibration in the IR spectrum.[29] In a medium of hexane and acetic acid a distinction between the isomerization and the fission route should be possible, because only the diazo compound reacts very rapidly with acetic acid. From the ratio of benzyl acetate and phenylheptane, produced by insertion of the carbene **54** into the CH bonds of the hexane, both processes display approximately the same efficiency.[29] Electron-releasing substituents in the phenyl ring favor isomerization. Additionally, a further transient species was detected, probably a 7aH-indazole **56**.[29]

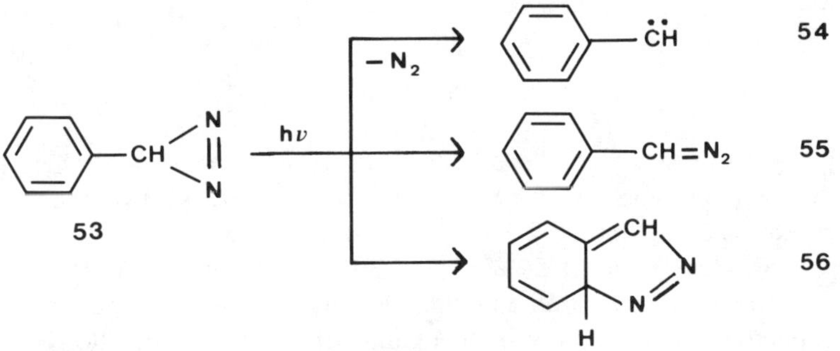

SCHEME 14.

A great deal of work has been done on the photochemistry of 3-chloro-3-methyldiazirine.[30-34] Irradiation into the long-wavelength transition at 325 nm leads to a quantum yield of 0.95 for the disappearance of the diazirine.[33] The observed reactions are summarized in Scheme 15.[34a] The dotted lines indicate processes, which play a minor or negligible role. This is of special interest for the indirect formation of the diazo compound: **58 + 57 → 58 + 60**.[33,35]

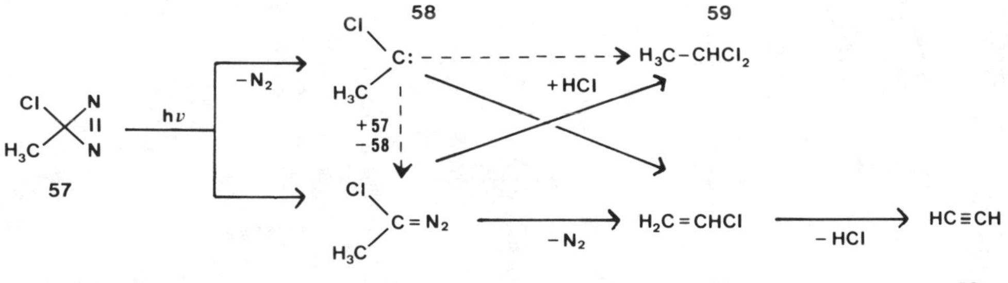

SCHEME 15.

A quantitative estimation for 3-phenyl-3-trifluoromethyldiazirine shows that irradiation at 350 nm yields 35% of the diazoisomer and 65% of the corresponding carbene.[36] Two distinct photochemical reaction paths have also been proved for 1,2-diazaspiro[2.5]oct-1-ene **63**.[37-39] The relative high percentage of azine generation serves as a hint for the formation of the diazo compound. A definite proof is provided by the IR spectroscopic detection[38] of the diazo-N,N-stretching vibration at ~ 2060 cm^{-1} and by the trapping reaction[39] with dimethyl acetylendicarboxylate in a 1,3-dipolar cycloaddition.

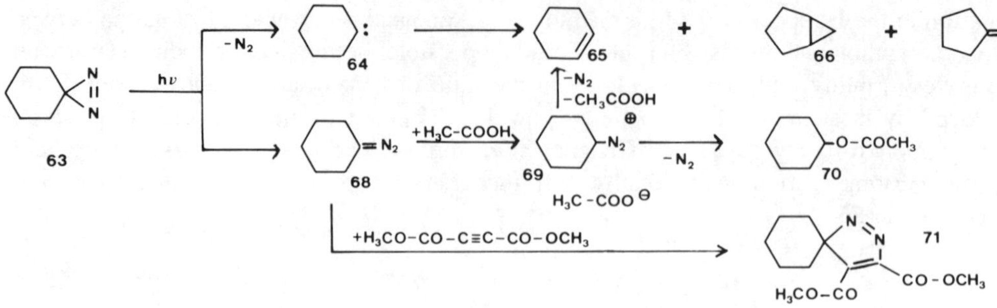

SCHEME 16.

Protonation of **68** by acetic acid leads to the diazonium structure **69**, which furnishes the acetate **70** and cyclohexene **65**. By using [²H] acetic acid, 33% of the cyclohexene is monodeuteriated. The remaining 67% does not contain deuterium because it originates in carbene **64**. The overall yield of 59% **65** and 39% **70** suggests that 60% of the reacting diazirine molecules isomerize to **68** and 40% directly eliminate nitrogen.[38] Finally, in one example, the photoreaction of 3-butyl-3-phenyldiazirine **72**,[40] isomerization is the only route. In this case the carbene formation can be ruled out because the olefin **75** is completely deuteriated in the presence of [²H₄] acetic acid. *m*-Chloroperbenzoic acid yields the oxidation products valerophenone **76** and 2-phenyl-3-propyloxirane **77** — a benzoate is not formed, probably due to steric reasons. In a Norrish type II rearrangement **76** is transformed to acetophenone **78**.[40]

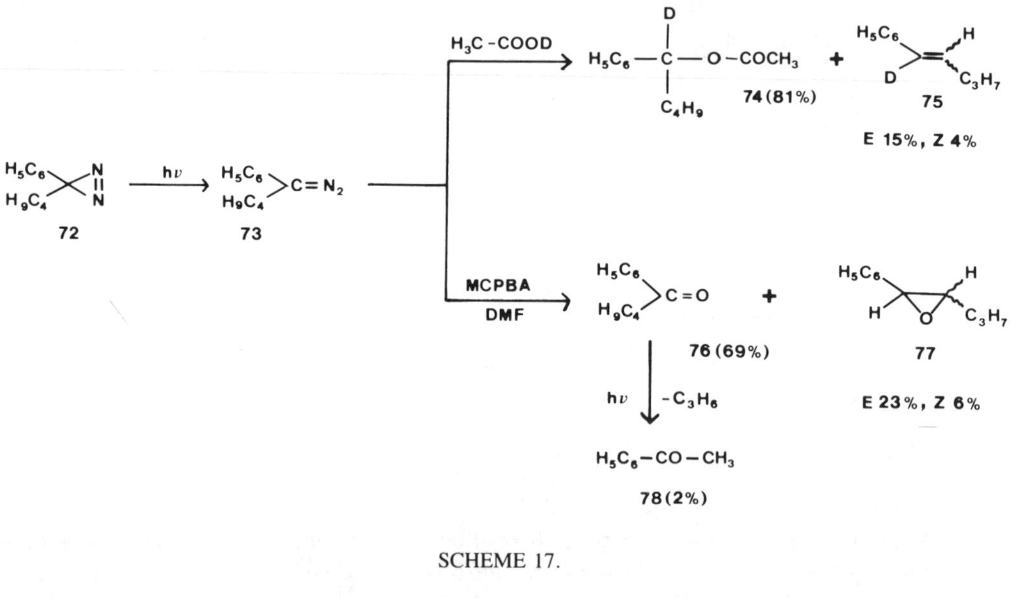

SCHEME 17.

Practically the same results are obtained in the thermal decomposition of **72**, i.e., the thermal process also proceeds by a one-bond rupture of the diazirine.[40,41] The investigation of the photochemical decomposition of 3-chloro-3-benzyldiazirine in acetic acid supports the carbenic mechanism.[41a] Since halodiazomethanes decompose at $-60°C$, it is unlikely that they participate in the product-forming steps.

Scheme 13 is also valid for the thermolysis of diazirines, and again the partition between the isomerization and the fission route depends on the particular system. An example demonstrating the opposite behavior of **72** is perfluoro-3-acetyl-3-methyldiazirine **34**, that decomposes at 80°C in solution or in the vapor phase by direct elimination of nitrogen. The corresponding diazoketone **33** is stable at much higher temperatures.[21]

Overberger and Anselme suggested the first assumption of a thermal ring opening of a diazirine to a diazo compound. They attempted to synthesize 3,3-diphenyldiazirine via the oxidation of 3,3-diphenyldiaziridine, but isolated only diphenyldiazomethane.[42] Since that time, many investigations have been performed on this topic.[26] In 1976, Jennings and Liu reported the first isolation of a diazoalkane in a diazirine thermolysis.[41,43] 3-Butyl-3-phenyldiazirine **72** thermally decomposes to *cis*- and *trans*-1-phenylpent-1-ene.

$$\Delta, \text{DMSO}, \ 100.2°$$

$$72 \longrightarrow 73 \longrightarrow N_2 + H_5C_6-\ddot{C}-C_4H_9$$

$$79$$

$$k_1 = 6.75 \times 10^{-4} \ s^{-1} \qquad k_2 = 2.23 \times 10^{-4} \ s^{-1}$$

$$A = 10^{13.26} \ s^{-1} \qquad A = 10^{12.03} \ s^{-1}$$

$$E_a = 28.08 \ \text{kcal/mol} \qquad E_a = 26.81 \ \text{kcal/mol}$$

SCHEME 18.

The disappearance of **72** (measured by the UV absorption) is approximately three times faster than the nitrogen evolution. Diazo compound **73** can be detected spectroscopically, isolated, and shown to account quantitatively for the nitrogen produced. Thermolysis in acetic acid gives practically the same rate constants for the disappearance of **72** and the formation of N_2, because the acid-catalyzed decomposition of diazo compounds is very rapid.[41]

The isomerization of 3-methyl-3-phenyldiazirine to 1-phenyldiazoethane is also first order and probably unimolecular, but the kinetics for the subsequent reactions are complicated by several competing processes.[43,44]

Kinetic investigations have also been made for the process **19** $\rightarrow$ **18**[12] at room temperature.

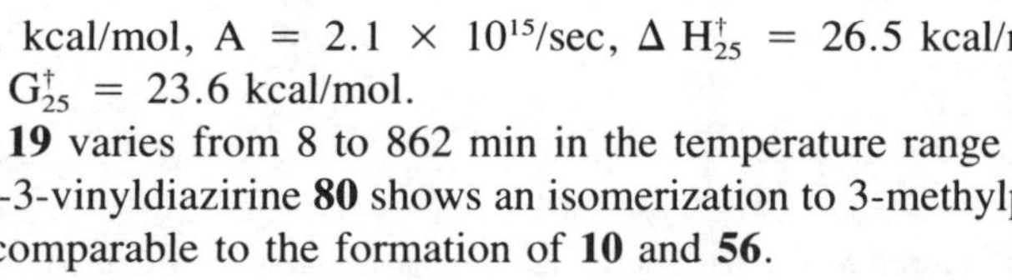

SCHEME 19.

where $E_a = 27.1$ kcal/mol, $A = 2.1 \times 10^{15}$/sec, $\Delta H_{25}^{\ddagger} = 26.5$ kcal/mol, $\Delta S_{25}^{\ddagger} = 9.7$ cal/K·mol, and $\Delta G_{25}^{\ddagger} = 23.6$ kcal/mol.

The halflife of **19** varies from 8 to 862 min in the temperature range between 53.4 and 18.5°C. 3-Methyl-3-vinyldiazirine **80** shows an isomerization to 3-methylpyrazole **83**.[45] The step **81** $\rightarrow$ **82** is comparable to the formation of **10** and **56**.

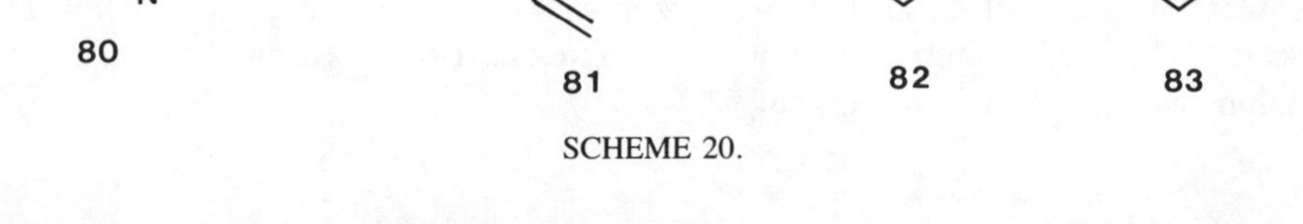

SCHEME 20.

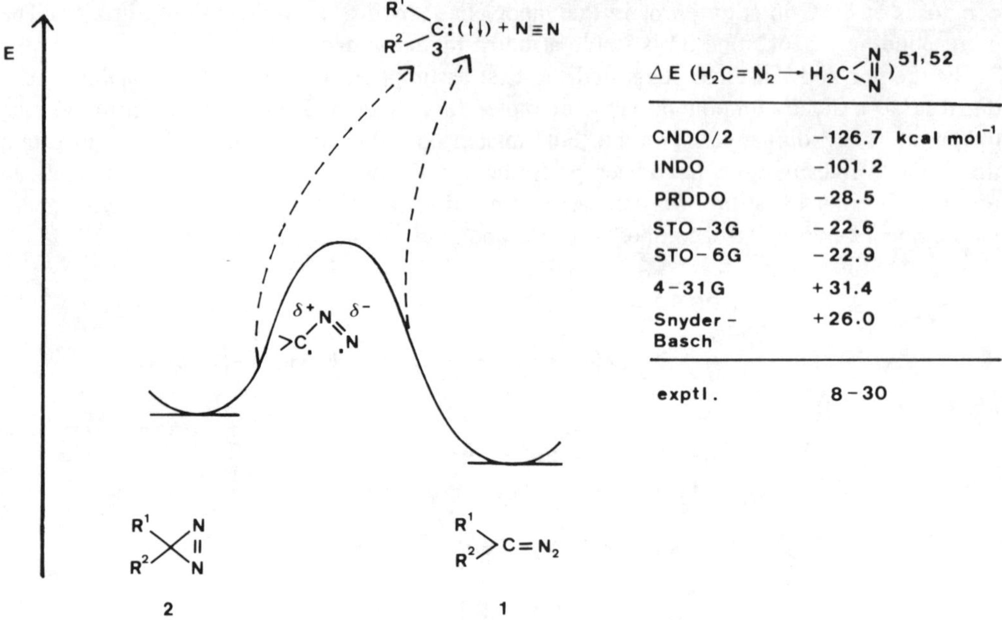

FIGURE 3.   Schematic diagram of the thermal isomerization of diazirines **2** and diazoalkanes **1**.

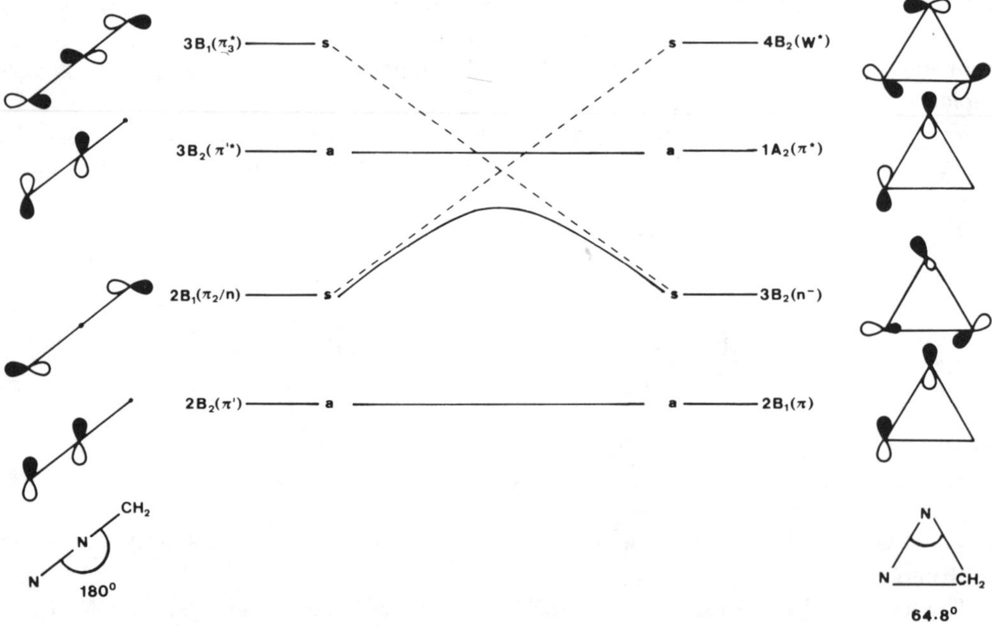

FIGURE 4.   MO correlation diagram for the interconversion of diazomethane and diazirine.

IV. MECHANISTIC ASPECTS

The interconversion of diazirines and diazoalkanes is an extremely interesting process because it is the smallest representation of a ring-chain isomerization. Several theoretical publications have appeared on this topic.[46-50a]

Diazirine and diazomethane correlate in the ground state as different minima of the S_0-energy-hypersurface CH_2N_2 (Figure 3). Unfortunately, there is an uncertainty about the reaction energy of this isomerization.[51,52a] However, the ring opening of diazirine is most likely an exothermic process with a $\Delta H°$ value of about -20 to -30 kcal/mol.

The transition state on the reaction coordinate $2 \rightarrow 1$ should be reached by a homolytic cleavage of a CN bond. It is reasonable to assume a singlet diradical somewhat contaminated by an ionic character.[48] The crucial question is, which crossroads exist leading to a minimum at the "edge" of the energy hypersurface with an infinite distance between the singlet carbene fragment and the nitrogen molecule? The ratio of isomerization and fission is determined by the relative heights of the saddle points (transition states) on the reaction coordinates. The Arrhenius activation energy E_a for the thermal decomposition of diazirines lies in the range of 27 to 33 kcal/mol.[53] The E_a value for the isomerization is about 27 to 30 kcal/mol.

The $\Delta H°$ of Figure 3 ensures an extremely low probability for the thermal isomerization in the direction of $1 \rightarrow 2$. On the contrary, photochemical isomerizations are known in both directions.

Figure 4 shows a MO correlation diagram for the interconversion.[12,48] The 180° angle in diazomethane is bent along the reaction coordinate to 64.8°, the angle in diazirine. The bisecting plane of the CH_2 group is the only symmetry element conserved throughout the entire process. The π and π^* orbitals perpendicular to this plane are marginally affected. A strong perturbation must be assumed for the in-plane MOs having n or Walsh (W) character. In a "natural correlation"[48] in response to the phase continuity rule, π_2 ($2B_1$) tends to correlate with the antibonding W^* ($4B_2$) and π_3^* ($3B_1$) with n^- ($3B_2$). Because of the same symmetry and multiplicity we encounter an avoided crossing.

The state correlation diagram, easily deduced from Figure 4 confirms the correlation of the first excited singlet states of diazirines **2** and diazomethanes **1**.[12,48] On the basis of the calculated potential energy curves[48] the photoisomerization $2 \rightarrow 1$ is a "downhill" process. The reverse process $1 \rightarrow 2$ is more problematic from the energetical point of view. A crossing of the triplet $\pi\pi^*$ surface offers a possible reaction route. However, a surface switching $S_1 \rightarrow T_1 \rightarrow S_0$ is not straightforward. It is uncertain how a funnel from S_1 to the ground state energy hypersurface can be reached. Concerning the triplet participation, a few important experiments are still lacking. Analogous to the thermal route, the photochemical isomerization is confined by the nitrogen extrusion leading to a singlet carbene.[12,48] The formation of diazoalkanes from diazirines via the carbene fragment plays a minor role.[6] Recently SINDO1, a semiempirical MO method, was applied for the enlightenment of the photochemical reaction paths of diazirine, 3,3-dimethyldiazirine, and 3-formyldiazirine.[50a] The results differ significantly from the discussion above, because low-lying $^1n\sigma^*$- and also $^3n\sigma^*$-states are relevant in this calculation.

A main point in the discussion of mechanistic aspects concerns the type of excitation that might be achieved in the reaction products. Although adiabatic photoreactions leading to electronically excited singlet or triplet states are less likely for such isomerization processes, this possibility should be considered carefully for the carbene formation. Excited singlet carbenes, proposed for the photochemical Wolff rearrangement,[13] have been claimed recently for the decomposition of diazirines.[54] Of course, vibrational excitation and energy partitioning are further important features in this process.

Certainly, various additional examples of direct *diazirine-diazoalkane interconversions* will be discovered in future investigations. The crucial point of approach is to suppress the nitrogen elimination on both sides of the valence isomerization. In the past, suitable conditions may have been lacking. For example, in the competition of the isomerization and the fission route, one process may surpass the other totally, depending on the particular system. The photolysis ($\lambda \geq 340$ nm) of 3-chloro-3-methoxydiazirine in an argon matrix of 10 K did not yield, for example, any trace of the corresponding diazo compound.[55]

Numerous experimental as well as theoretical questions concerning the interconversion of diazoalkanes and diazirines remain unanswered.

REFERENCES

1. **Angeli, A.**, *Atti Reale Acad. Lincei*, 16(II), 790, 1907; 20(I), 626, 1911.
2. **Thiele, J.**, Über die Konstitution der aliphatischen Diazoverbindungen und der Stickstoffwasserstoffsäure, *Ber. Dtsch. Chem. Ges.*, 44, 2522, 1911.
3. **Langmuir, I.**, Isomorphism, isosterism and covalence, *J. Am. Chem. Soc.*, 41, 1543, 1919.
4. **Curtius, T.**, Über die Konstitution der fetten Diazo- und Azokörper und über die Bildung des Diamids und seiner Derivate, *J. Prakt. Chem.*, 39, 107, 1889.
5. **von Pechmann, H.**, Über Diazomethan, *Ber. Dtsch. Chem. Ges.*, 27, 1888, 1894.
6. **Moore, C. B. and Pimentel, G. C.**, Matrix reaction of methylene with nitrogen to form diazomethane, *J. Chem. Phys.*, 41, 3504, 1964; **Shilov, A. E., Shteinman, A. A., and Tjabin, M. B.**, Reaction of carbenes with molecular nitrogen, *Tetrahedron Lett.*, 4177, 1968.
7. **Schulz, R. and Schweig, A.**, Existenz von 1,2,3-Benzooxadiazol in der Gasphase, *Angew. Chem.*, 91, 737, 1979; *Angew. Chem. Int. Ed. Engl.*, 18, 692, 1979.
8. **Schulz, R. and Schweig, A.**, 1,2,3-Benzooxadiazol — Nachweis in Argon-Matrix und in Lösung, *Angew. Chem.*, 96, 494, 1984; *Angew. Chem. Int. Ed. Engl.*, 23, 509, 1984.
9. **Lowe, G. and Parker, J.**, Photochemical isomerization of some diazo-compounds into diazirines, *J. Chem. Soc. Chem. Commun.*, 1135, 1971.
10. **Franich, R. A., Löwe, G., and Parker, J.**, Photochemical interconversion of some diazo-amides and diazirinecarboxamides, *J. Chem. Soc. Perkin Trans. I*, 2034, 1972.
11. **Voigt, E. and Meier, H.**, Diazirine als photochrome Valenzisomerie von α-Diazocarbonylverbindungen, *Angew. Chem.*, 87, 109, 1975; *Angew. Chem. Int. Ed. Engl.*, 14, 103, 1975.
12. **Voigt, E. and Meier, H.**, Über die Valenzisomerie von Diazoverbindungen und Diazirinen, *Chem. Ber.*, 108, 3326, 1975.
13. **Meier, H. and Zeller, K.-P.**, Die Wolff-Umlagerung von α-Diazocarbonyl-Verbindungen, *Angew. Chem.*, 87, 52, 1975; *Angew. Chem. Int. Ed. Engl.*, 14, 32, 1975.
14. **Rando, R. R.**, Conformational and solvent effects on carbene reactions, *J. Am. Chem. Soc.*, 92, 6706, 1970.
15. **Moll, F.**, Condensed 2-azetidinones, *Arch. Pharmacol.*, 301, 320, 1968; **Brunwin, D. M., Lowe, G., and Parker, J.**, The total synthesis of a nuclear analogue of the penicillin-cephalosporin antibiotics, *J. Chem. Soc. Ser. C*, 3756, 1971; **Earle, R. H., Hurst, D. T., and Viney, M.**, Synthesis and hydrolysis of some fused-ring-β-lactons, *J. Chem. Soc. Ser. C*, 2093, 1969; **Corey, E. J. and Felix, A. M.**, A new synthetic approach to the penicillins, *J. Am. Chem. Soc.*, 87, 2518, 1965.
16. **Moriconi, E. J. and Murrey, J. J.**, Pyrolysis and photolysis of 1-methyl-3-diazooxindole. Base decomposition of isatin 2-tosyl-hydrazone, *J. Org. Chem.*, 29, 3577, 1964.
17. **Srinivasan, R.**, Photochemistry of methoxyacetone, *J. Am. Chem. Soc.*, 84, 2475, 1962.
18. **Livinghouse, T. and Stevens, R. V.**, A direct synthesis of spiro-activated cyclopropanes from alkenes via the irradiation of isopropylidene diazomalonate, *J. Am. Chem. Soc.*, 100, 6479, 1978.
19. **Miyashi, T., Nakajo, T., and Mukai, T.**, Reversible photochemical valence isomerization between α-diazo-ketones and α-keto-diazirines, *J. Chem. Soc. Chem. Commun.*, 442, 1978.
20. **Schmitz, E., Stark, A., and Hörig, C.**, Ein Diazoketon mit Dreiringstruktur, *Chem. Ber.*, 98, 2509, 1965.
21. **Laganis, E. D., Janik, D. S., Curphey, T. J., and Lemal, D. M.**, Photochemistry of perfluoro-3-diazo-2-butanone, *J. Am. Chem. Soc.*, 105, 7457, 1983.
22. **Chapman, O. L., Chang, C.-C., Kolc, J., Jung, M. E., Lowe, J. A., Barton, T. J., and Tumey, M. L.**, 1,1,2-Trimethylsilaethylene, *J. Am. Chem. Soc.*, 98, 7844, 1976.
23. **Chedekel, M. R., Skoglund, M., Kreeger, R. L., and Shechter, H.**, Solid state chemistry, discrete trimethylsilylmethylene, *J. Am. Chem. Soc.*, 98, 7846, 1976.
24. **Ando, W., Tanikawa, H., and Sekiguchi, A.**, Ring expansion of silacyclopentadienylcarbene to sila-benzene, *Tetrahedron Lett.*, 24, 4245, 1983.

25. **Erni, B. and Khorana, H. G.,** Fatty acids containing photoactivable carbene precursors. Synthesis and photochemical properties of 3,3-bis(1,1-difluorohexyl)diazirine and 3-(1,1-difluorooctyl)-3 H-diazirine, *J. Am. Chem. Soc.,* 102, 3888, 1980.
26. Reviews on photochemistry and thermochemistry of diazirines: **Liu, M. T. H.,** The thermolysis and photolysis of diazirines, *Chem. Soc. Rev.,* 11, 127, 1982; **Frey, H. M.,** Photolysis of the diazirines, *Pure Appl. Chem.,* 9, 527, 1964; The photolysis of the diazirines, *Adv. Photochem.,* 4, 225, 1966; **Schmitz, E.,** Dreiringe mit zwei Heteroatomen, Springer Verlag, Basel, 1967.
27. **Amrich, M. J. and Bell, J. A.,** Photoisomerization of diazirine, *J. Am. Chem. Soc.,* 86, 292, 1964.
28. **Lahmani, F.,** Reversibility of the pressure-induced intersystem crossing in methylene, *J. Phys. Chem.,* 80, 2623, 1976.
29. **Smith, R. A. G. and Knowles, J. R.,** The preparation and photolysis of 3-aryl-3H-diazirines, *J. Chem. Soc. Perkin Trans. II,* 686, 1975.
30. **Moss, R. A. and Mamantov, K.,** Methylchlorocarbene, *J. Am. Chem. Soc.,* 92, 6951, 1970.
31. **Cadman, P., Engelbrecht, W. J., Lotz, S., and Van der Merwe, S. W. J.,** Energy distribution in reaction products. Photolysis of 3-halo-3-methyldiazirines, *J. S. Afr. Chem. Inst.,* 27, 149, 1974; *Chem. Abstr.,* 82, 178111t, 1975.
32. **Jones, W. E., Wasson, J. S., and Liu, M. T. H.,** Photolysis of 3-chloro-3-methyldiazirine, *J. Photochem.,* 5, 311, 1976.
33. **Frey, H. M. and Penny, D. E.,** Photolysis of 3-methyl-3-chlorodiazirine, *J. Chem. Soc. Faraday Trans. I,* 73, 2010, 1977.
34. **Figuera, J. M., Pérez, J. M., and Tobar, A.,** Energy distribution and mechanism in 3-chloro-3-methyldiazirine photolysis, *J. Chem. Soc. Faraday Trans. I,* 74, 809, 1978.
34a. See also **Liu, M. T. H.,** Energy barrier for 1,2-hydrogen migration in benzylchlorocarbene, *J. Chem. Soc. Chem. Commun.,* 982, 1985.
35. Compare also Reference 28.
36. **Brunner, J., Senn, H., and Richards, F. M.,** 3-Trifluoromethyl-3-phenyldiazirine, *J. Biol. Chem.,* 255, 3313, 1980; *Chem. Abstr.,* 93, 3412s, 1980.
37. **Frey, H. M. and Stevens, I. D. R.,** Reactions of vibrationally excited molecules. III. The photolysis of cyclohexanespiro-3-diazirine, *J. Chem. Soc.,* 4700, 1964.
38. **Bradley, G. F., Evans, W. B. L., and Stevens, I. D. R.,** Thermal and photochemical decomposition of cycloalkanespirodiazirines, *J. Chem. Soc. Perkin Trans. II,* 1214, 1977.
39. **Gstach, H. and Kisch, H.,** Synthese von Spiroverbindungen durch Photoisomerisierung von Spirodiazirinen in Gegenwart von Alkinen, *Chem. Ber.,* 115, 2586, 1982.
40. **Liu, M. T. H., Palmer, G. E., and Chishti, N. H.,** Photolysis and thermolysis of 3-n-butyl-3-phenyldiazirine, *J. Chem. Soc. Perkin Trans. II,* 53, 1981.
41. **Jennings, B. M. and Liu, M. T. H.,** Mechanism of thermal decomposition of diazirine. Evidence for diazomethane intermediate, *J. Am. Chem. Soc.,* 98, 6416, 1976.
41a. **Liu, M. T. H., Chishti, N. H., Tencer, M., Tomioka, H., and Izawa, Y.,** Reactive intermediates in the photolysis and thermolysis of 3-chloro-3-benzyldiazirine, *Tetrahedron,* 40, 887, 1984: compare also **Liu, M. T. H. and Subramanian, R.,** Reaction of benzychlorocarbene with hydrogen chloride, *J. Org. Chem.,* 50, 3218, 1985.
42. **Overberger, C. G. and Anselme, J. P.,** Novel rearrangement: ring opening of a cyclic diazo compound to a diazoalkane, *Tetrahedron Lett.,* 1405, 1963.
43. **Liu, M. T. H. and Jennings, B. M.,** Consecutive reactions in the thermal decomposition of phenylalkyldiazirines, *Can. J. Chem.,* 55, 3596, 1977.
44. **Liu, M. T. H. and Ramakrishnan, K.,** Thermal decomposition of phenylmethyldiazirine. Effect of solvent on product distribution, *J. Org. Chem.,* 42, 3450, 1977.
45. **Liu, M. T. H. and Toriyama, K.,** Thermal unimolecular isomerization of 3-methyl-3-vinyldiazirine, *Can. J. Chem.,* 51, 2393, 1973.
46. **Hoffmann, R.,** Extended Hückel theory. VI. Excited states and photochemistry of diazirines and diazomethanes, *Tetrahedron,* 22, 539, 1966.
47. **Snyder, J. P., Boyd, R. J., and Whitehead, M. A.,** Chelotropic reactions: the thermal distribution of diazirine, *Tetrahedron Lett.,* 4347, 1972.
48. **Bigot, B., Ponec, R., Sevin, A., and Devaquet, A.,** Theoretical ab initio SCF investigation of the photochemical behavior of three-membered rings. I. Diazirine, *J. Am. Chem. Soc.,* 100, 6575, 1978.
49. **Bigot, B., Sevin, A., and Devaquet, A.,** Natural MO and state correlation. A governing concept to predict the profile of the potential energy curves. Application to the diazomethane-diazirine system, *Proc. 7th IUPAC Symp. Photochemistry,* 1978, 46.
50. **Pérez, J. M.,** Evidence for the participation of an isomerization pathway in diazirine photolysis, *J. Chem. Soc. Faraday Trans. I.,* 78, 3509, 1982.
50a. **Müller-Remmers, P. L. and Jug, K.,** SINDO 1 study of photochemical reaction mechanisms of diazirines, *J. Am. Chem. Soc.,* 107, 7275, 1985.

51. **Halgren, T. A., Kleier, D. A., Hall, J. H., Jr., Brown, L. D., and Lipscomb, W. N.,** Speed and accuracy in molecular orbital calculations. A comparison of CNDO/2, INDO, PRDDO, STO-3G, and other methods, including AAMOM, VRDDO, and ESE MO, *J. Am. Chem. Soc.,* 100, 6595, 1978.
52. **Bell, J. A.,** On mass spectra and appearance potentials of diazirine and diazomethane, *J. Chem. Phys.,* 41, 2556, 1964.
53. **Engel, P. S.,** Mechanism of the thermal photochemical decomposition of azoalkanes, *Chem. Rev.,* 80, 99, 1980.
54. **Warner, P. M.,** On carbene alkene complexes, *Tetrahedron Lett.,* 4211, 1984.
55. **Sheridan, R. S. and Kasselmayer, M. A.,** Infrared spectrum and photochemistry of methoxychlorocarbene, *J. Am. Chem. Soc.,* 106, 436, 1984.

Chapter 7

DISSOCIATION ENERGETICS OF SIMPLE DIAZIRINES BY PHOTON AND ELECTRON IMPACT

Hideo Okabe

TABLE OF CONTENTS

I. INTRODUCTION

Diazirines were first synthesized in 1961 by Schmitz and Ohme[1] and subsequently by Graham.[2] Unlike their isomeric linear diazo compounds, diazirines are remarkably unreactive with alkalines and strong acids, although they are thermally unstable and tend to explode in a condensation vaporization cycle at liquid nitrogen temperature. Because of the diazirine explosive property, their heats of formation have not been directly measured by a calorimetric method. Instead, the heat of formation has been derived either from appearance potential of a fragment ion,[3] or from the appearance threshold of fluorescence via electronically excited radicals.[4] Diazirines are used as a source of carbenes, the simplest being methylene. The reactivities of methylene are remarkably different, depending on whether the methylene is in the singlet or the ground triplet state. The singlet CH_2 is more reactive than the triplet. The singlet adds to the double bond of olefins, forming stereospecific cyclopropane products of respective olefins.[5] On the other hand, the reaction of more complex carbenes, such as ethylidene CH_3HC from 3-methyldiazirine photolysis, has not been conclusively established. The rearrangement of ethylidene to ethylene by H atom transfer must be rapid. Reaction products of ethylidene with large excesses of ethylene, propylene, and allene have not been found. However, Cl substituted ethylidene, CH_3ClC, has recently been shown to add stereospecifically to *cis*- and *trans*-butene to form respective 1-chloro-1-methylcyclopropanes.[25] Vinyl chloride in comparable yield has also been observed. The results show that the H atom migration in CH_3ClC becomes sufficiently slow to compete with the addition reaction of the carbene to the double bond.

The energetics of the photolysis of diazirines, that is, the total available energy after photolysis and the distribution of the energy in each photofragment, have not been well-established, because (1) the heats of formation of diazirines have not been obtained directly by a calorimetric method. Rather, they are derived from the threshold of fragment ion formation or electronically excited fragment production, and (2) the internal and translational energies in photofragments have not been directly measured. In the photolysis of 3-chloro-3-methyldiazirine, it has been estimated from the effect of collisions on the product distribution that about 70% of the total available energy (the photon energy plus the exothermicity of the reaction) is imparted to the carbene fragment.

The following discussion describes the photon and electron impact studies of six simple diazirines, namely, diazirine, methyldiazirine, dimethyldiazirine, 3-chloro-3-methyldiazirine, 3-bromo-3-methyldiazirine, and difluorodiazirine. Frey[7] and more recently Liu[8] have reviewed the photolysis and thermolysis of more complex diazirines.

II. PHOTON AND ELECTRON IMPACT STUDIES OF DIAZIRINES

A. Diazirine

Diazirine, a cyclic diazomethane, belongs to a C_{2v} symmetry group, where the C_2 axis is chosen along the Z axis and the N=N bond parallels the X direction. The CH_2 group lies in the YZ plane.[9] The molecule is strongly asymmetric. The O–O band lies at 323.0 nm.[10] The UV spectrum in the 200 to 350 nm region is shown in Figure 1.[2] The molar absorption coefficient, ϵ, at 308.5 nm is 176 M^{-1} cm^{-1}. The spectrum shows many sharp regularly spaced peaks characterized by the strong v_3 CN symmetric stretch with 797 cm^{-1} intervals and another progression with 848 cm^{-1} intervals which is probably a combination of v_5 (torsional oscillation about the Z axis) and v_9 (asymmetric CN stretching vibration).[10] The bands are diffuse under high resolution.[11]

The heat of formation of diazirine has been derived from electron impact[3] and photon impact[4] studies. The value from electron impact is obtained via the appearance potential of the CH_2^+ ion. If the CH_2^+ ion is formed in the ground state,[3] ΔH_f° (diazirine) is 79 kcal/mol,

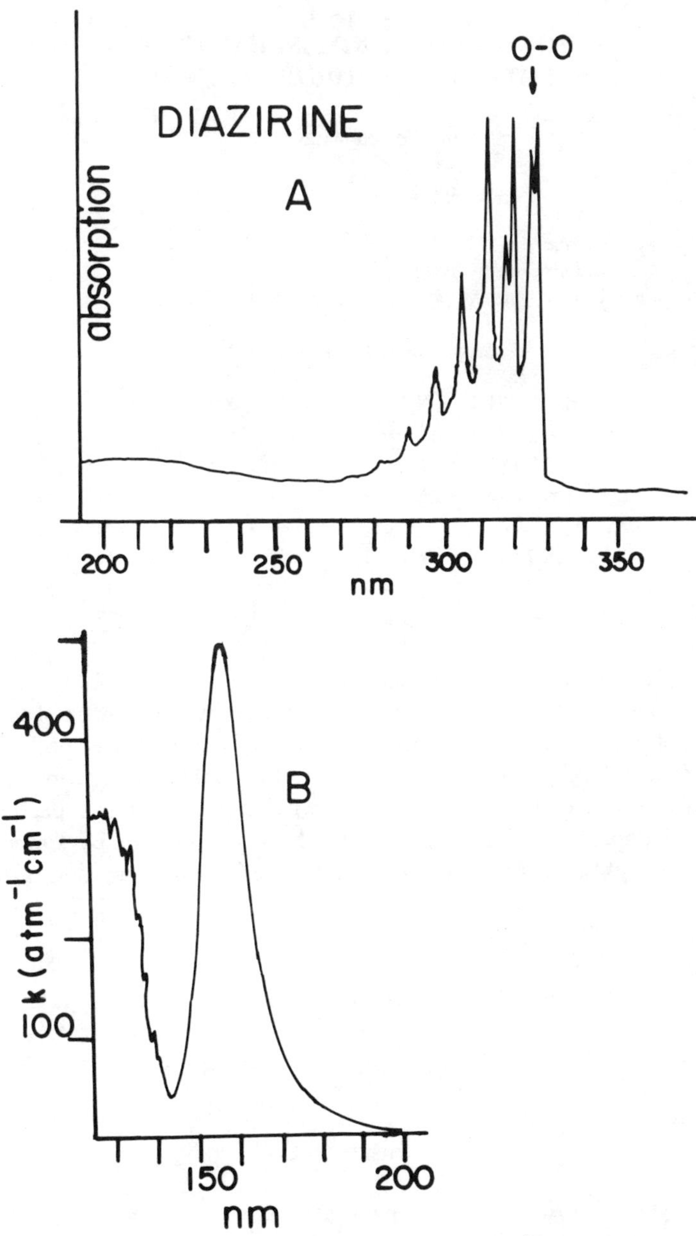

FIGURE 1.   (A) The UV spectrum of diazirine. The O–O band is at 323.0 nm.[10] The molar absorption coefficient at 308.5 nm is 176 dm³/mol/cm. (Reprinted with permission from Graham, W. H., *J. Am. Chem. Soc.*, 84, 1063. Copyright 1962 American Chemical Society.) (B) The vacuum UV spectrum of diazirine. The absorption coefficient is in units of atm⁻¹ cm⁻¹, base e. (Reprinted with permission from Laufer, A. H. and Okabe, H., *J. Phys. Chem.*, 76, 3504. Copyright 1972 American Chemical Society.)

using an appearance potential of 11.0 ± 0.1 eV and $\Delta H_f^{\circ}(CH_2^+)$[12] of 332 kcal/mol. On the other hand, Bell[13] argued that the heat of formation could be 101 kcal/mol if the CH_2^+ ion is formed in the electronically excited state. The ionization and appearance potentials of diazirines from which the heats of formation are derived are given in Table 1. The heat of formation can also be derived from the appearance threshold of $CH(A \rightarrow X)$ emission on the

Table 1
IONIZATION AND APPEARANCE
POTENTIALS FOR DIAZIRINES

Ionization Potentials (I.P.)

Compound	eV	Ref.
Diazirine	10.18 ± 0.05	3
3-Chloro-3-methyldiazirine	$<8.4 \pm 0.1$	32
3-Bromo-3-methyldiazirine	$<8.4 \pm 0.1$	32

Appearance Potentials (A.P.)

	eV	Ref.
Diazirine $\xrightarrow{e}$ $CH_2^+ + N_2 + 2e$	11.0 ± 0.1	3
Diazirine $\xrightarrow{e}$ $CHN_2^+ + H + 2e$	14.2 ± 0.1	3
CMD $\xrightarrow{e}$ $C_2H_3Cl^+ + N_2 + 2e$	8.4 ± 0.1	32
$\xrightarrow{e}$ $C_2H_3N_2^+ + Cl + 2e$	11.0 ± 0.1	32
$\xrightarrow{e}$ $C_2H_3^+ + N_2 + Cl + 2e$	10.4 ± 0.1	32
BMD $\xrightarrow{e}$ $C_2H_3Br^+ + N_2 + 2e$	8.4 ± 0.1	32
$\xrightarrow{e}$ $C_2H_3N_2^+ + Br + 2e$	10.3 ± 0.1	32
$\xrightarrow{e}$ $C_2H_3^+ + N_2 + Br + 2e$	9.8 ± 0.1	32

vacuum UV (VUV) photolysis of diazirine.[4] The value of 61 kcal/mol is obtained from the VUV photolysis. The upper limit of 66 kcal/mol is placed by the absence of the CH emission at 147 nm excitation. The electron impact value of 79 kcal/mol may be less reliable than the photon impact value, because the appearance threshold of fragment ions in electron impact is usually less precise than that of excited radicals in photon impact (see Reference 5, p. 102). The independent measurement of the heat of formation by a more direct calorimetric method is highly desirable. In this chapter the heat of formation of 61 kcal/mol is recommended.

1. Photolysis of Diazirine

The photolysis of diazirine at 313 nm produces ethylene and nitrogen.[14]

$$\text{Diazirine} \xrightarrow{h\nu} CH_2 + N_2 \tag{1}$$

$$CH_2 + \text{Diazirine} \rightarrow C_2H_4 + N_2 \tag{2}$$

The excess available energy for the fragments in Reaction 1 is 50 kcal/mol, which is the photon energy of 91 kcal/mol minus the heat of reaction of 41 kcal/mol for Reaction 1, assuming the singlet CH_2 is formed. The quantum yield of disappearance of diazirine[15] is 2.0 ± 0.5 at 320 nm. The methylene radical initially formed is probably singlet, because the addition to *trans*-butene-2 and 2-methylbutene-2 is stereospecific. Since methylene from diazirine is more selective toward primary and secondary C–H bond insertion, Frey and Stevens[14] have concluded that the CH_2 radicals from diazirine contain more vibrational energy than those from diazomethane or ketene, although direct proof of the hypothesis is lacking. Amrich and Bell reported[15] that diazomethane is produced in diazirine photolysis, whereas Moore and Pimentel[16] could not detect any diazomethane. Further work is needed to resolve this discrepancy.

B. 3-Methyldiazirine (MD)

The absorption spectrum[7] of MD is shown in Figure 2. The molar absorption coefficient[17]

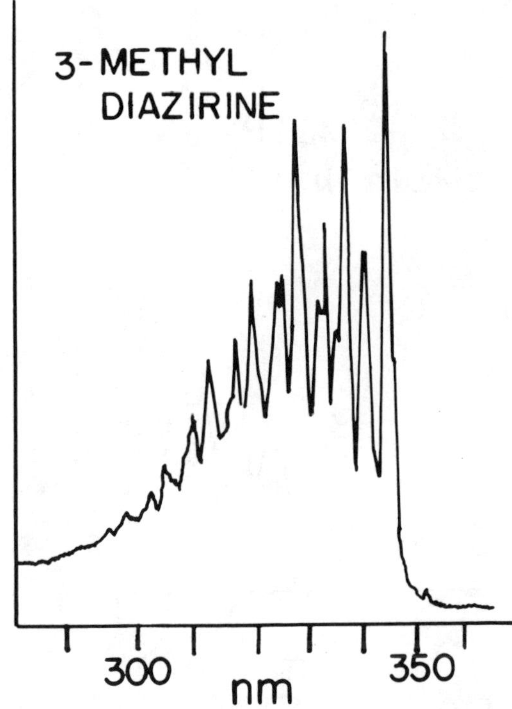

FIGURE 2.   The UV absorption spectrum of MD. The band extends from 294 to 350 nm; it consists of many sharp lines with the maximum at 341 nm, where the absorption coefficient is 470 dm^3/mol/cm. (From Frey, H. M., in *Advances in Photochemistry,* Vol. 4, Noyes, W. A., Jr., Hammond, G. S., and Pitts, J. N., Eds., John Wiley & Sons, New York, 1966, 225. With permission.)

at 341 nm is about 470 M^{-1} cm^{-1}. The heat of formation of MD apparently has not been obtained. The photolysis of MD at 313 nm produced ethylene, acetylene, nitrogen, and hydrogen.[17] These products can be explained by

$$MD \xrightarrow{h\nu} CH_3HC: + N_2 \tag{3}$$

$$CH_3HC: \rightarrow C_2H_4^\dagger \tag{4}$$

$$C_2H_4^\dagger \rightarrow C_2H_2 + H_2 \tag{5}$$

$$C_2H_4^\dagger \xrightarrow{M} C_2H_4 \tag{6}$$

where $C_2H_4^\dagger$ indicates vibrationally excited ethylene. A part of $C_2H_4^\dagger$ must contain energy not sufficient to decompose, because not all $C_2H_4^\dagger$ molecules dissociate into $C_2H_2 + H_2$ even at low pressures. The C_2H_4 to C_2H_2 ratio increases with the pressure of MD. The calculated C_2H_4/C_2H_2 values as a function of pressure can be reproduced if one assumes a wide distribution of internal energies in C_2H_4. New products were not found in the photolysis of MD in the presence of large excesses of olefins, suggesting that CH_3HC rearranges rapidly to $C_2H_4^\dagger$, prior to reacting with olefins.[17] The isomerization of MD to diazoethane lacks evidence.

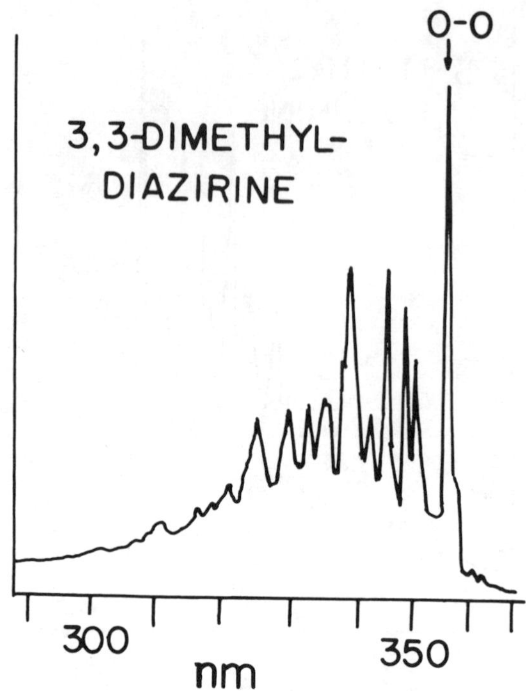

FIGURE 3.   The UV spectrum of DMD. The band lies in the 290 to 370 nm region. The molar absorption coefficient at 360 nm is about 180 dm³/mol/cm. The O–O band is at 358.0 nm. (From Frey, H. M. and Stevens, I. D. R., *J. Chem. Soc.*, 3514, 1963. With permission.)

C. 3,3-Dimethyldiazirine (DMD)

Like diazirine, dimethyldiazirine belongs to a C_{2v} symmetry group. The absorption spectrum, shown in Figure 3, consists of a series of sharp peaks in the 290 to 370 nm region. The spectrum is characteristic of the strained nitrogen-nitrogen double bond. The molar absorption coefficient[18] at 360 nm is about 180 dm³/mol/cm. The vibrational analysis indicates the presence of a strong ν_3' symmetric N=N stretch and ν_7' symmetric CH_3 rocking progression.[19] The O–O band is at 358.0 nm.

The photolysis at 313 nm produces equimolar amounts of propylene and nitrogen at pressures above about 180 torr. However, at pressures below 60 torr, main hydrocarbon products are ethane, ethylene, propane, butene-1, 2,3-dimethylbutane, 4-methylpentene-1, and hexa-1,5-diene. The propylene to nitrogen ratio falls below unity. These findings may be explained by[18]

$$DMD \xrightarrow{h\nu} (CH_3)_2C: + N_2 \tag{7}$$

$$(CH_3)_2C: \rightarrow CH_3CHCH_2^{\ddagger} \tag{8}$$

$$CH_3CHCH_2^{\ddagger} \xrightarrow{M} CH_3CHCH_2 \tag{9}$$

$$CH_3CHCH_2^{\ddagger} \rightarrow CH_2CHCH_2 + H \tag{10}$$

$$CH_3CHCH_2^{\ddagger} \rightarrow CH_2CH + CH_3 \tag{11}$$

$$CH_3CHCH_2^{\ddagger} \rightarrow C_2H_2 + CH_4 \tag{12}$$

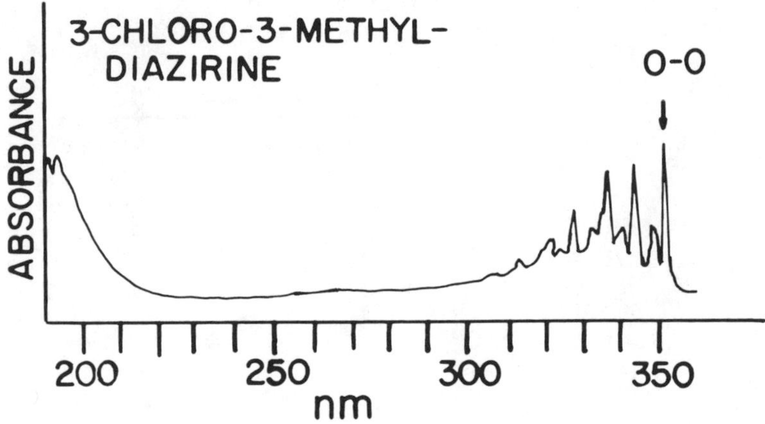

FIGURE 4.   The UV absorption spectrum of CMD. The UV spectrum lies in the 290 to 360 nm region. The molar absorption coefficient at 343 nm is about 100 dm³/ mol/cm. (From Cadman, P. et al., *J. S. Afr. Chem. Inst.*, 27, 149, 1974. With permission.)

and subsequent radical-radical reactions, where $CH_3CHCH_2^\dagger$ signifies vibrationally excited propylene. The initially formed dimethylcarbene rearranges to propylene via hydrogen atom transfer. The energy content of propylene in Reaction 8 is unknown, because (1) the heat of formation of DMD has not been measured and (2) the fraction of energy partitioned in propylene by the photolysis is uncertain. The addition of inert gas reduces the yields of hydrocarbons other than propylene.

The photolysis of DMD in cyclohexene liquid showed no addition product of dimethyl-carbene to the double bond. This indicates that the hydrogen shift in dimethylcarbene is more rapid than the addition to the double bond.

D. 3-Chloro-3-Methyldiazirine (CMD)

3-Chloro-3-methyldiazirine has C_s symmetry. The UV spectrum[20] of CMD is shown in Figure 4. Analysis of the spectrum indicates that it consists of two strong progressions[21] of 1454 and 802 cm^{-1}. The former is the v_3' symmetric NN stretch, the latter the v_8' symmetric CC stretch frequency. The O–O band is at 353.4 nm.

Results of the electron impact study are shown in Tables 1 and 2. The estimated heat of formation of CMD from the appearance potential[32] of $C_2H_3^+$ is 58 kcal/mol.

The photolysis products at 314, 334, and 353 nm are vinyl chloride, acetylene, hydrogen chloride, nitrogen, and about 3% 1,1-dichloroethane.[20,22-24] The ratio of C_2H_3Cl to C_2H_2 increased with the increase of inert gas pressure, such as SF_6[23] as shown in Figure 5. These results may be explained by

$$CMD \xrightarrow{h\nu} CH_3ClC: + N_2 \tag{13}$$

$$CH_3ClC: \rightarrow C_2H_3Cl^\dagger \tag{14}$$

$$C_2H_3Cl^\dagger \rightarrow C_2H_2 + HCl \tag{15}$$

$$C_2H_3Cl^\dagger \xrightarrow{M} C_2H_3Cl \tag{16}$$

where $C_2H_3Cl^\dagger$ indicates vibrationally excited C_2H_3Cl.

Table 2
MASS SPECTRA OF DIAZIRINES

		Relative intensity				
m/e	Ion	Diazirine[3,2]		DFD[30]	CMD[32]	BMD[32]
12	C^+	10.8	10.8	5.8		
13	CH^+	23.5	22.9			
14	CH_2^+	100.0	100.0		2.0	2.0
14	N^+			3.9		
15		1.9	2.2			
15	CH_3^+				5.0	3.0
25	C_2H^+				7.0	5.0
26	CN^+	2.3	1.8	3		
26	$C_2H_2^+$				37.0	29.0
27	CHN^+	6.6	5.8			
27	$C_2H_3^+$				78.0	100.0
28	N_2^+	8.5	17.8			
29	HN_2^+	1.9	2.3			
31	CF^+			63.6		
33	NF^+			2		
40	CN_2^+	2.1	1.8	1		6.0
41	CHN_2^+	10.7	10.8		9.0	14.0
42	$CH_2N_2^+$	48.1	43.1			
43		0.8	0.8			
45	CFN^+			4.7		
50	$CF_2{}^+$			100.0		
55	$C_2H_3N_2{}^+$				100.0	60.0
59	$CFN_2{}^+$			46.8		
62	$C_2H_3Cl^+$				29.0	
64	CF_2N^+			6.9		
64	$ClHN_2{}^+$				7.5	
69	$CF_3{}^+$			0.6		
75	$CClN_2{}^+$				7.0	
77					2.5	
78	$CF_2N_2{}^+$			0.1		
79	Br^+					6.0
81	Br^+					6.0
106	$C_2H_3Br^+$					9.0
108	$C_2H_3Br^+$					8.5
119	CN_2Br^+					3.0
121	CN_2Br^+					3.0

Figuera et al.[24] suggest that the pressure dependence curve such as the one shown in Figure 5 may be explained if two vibrationally excited states of vinyl chloride ($C_2H_3Cl^+$) are assumed, one dissociating immediately and the other with an internal energy of about 95 kcal/mol, corresponding to 70% of the total available energy of 143 kcal/mol (incident photon energy of 91 kcal/mol plus the exothermicity of 52 kcal/mol for CMD $\rightarrow$ C_2H_3Cl + N_2). The addition of HCl or HBr[22] to CMD produces 1,1-dihaloethane, suggesting the occurrence of insertion of carbene into HCl or HBr.[22]

$$CH_3ClC: + HX \rightarrow CH_3CHClX \qquad (17)$$

Frey and Penny[23] on the other hand, believe that 1,1-dichloroethane formed from carbene and HCl (Reaction 17) may be too "hot" to be stabilized by collisions, and consequently vinyl chloride and HCl would be formed. They suggest that chloromethyldiazomethane, formed by isomerization of CMD, reacts with HCl, forming 1,1-dichloroethane.

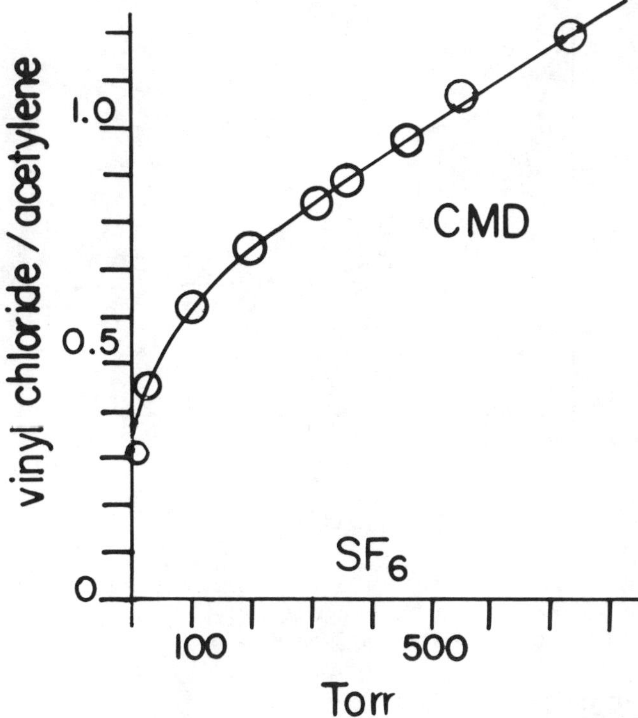

FIGURE 5.   Photolysis of CMD in the UV region. The ratio of C_2H_3Cl to C_2H_2 increases with an increase of SF_6. Initially formed vinyl chloride is vibrationally excited and it either decomposes to C_2H_2 and HCl or is stabilized by collisions with SF_6. (From Frey, H. M. and Penny, D. E., *J. Chem. Soc. Faraday Trans. 1*, 73, 2010, 1977. With permission.)

Moss and Mamantov[25] found that chloromethylcarbene formed from the photolysis of CMD adds stereospecifically to *cis-* and *trans-*butene to form respective 1-chloro-1-methylcyclopropanes. Vinyl chloride is also observed in comparable yield, suggesting that the stereospecific addition of chloromethylcarbene to the double bond competes with the isomerization of the carbene to vinyl chloride.

Becerra et al.[26] have found that the flash photolysis of CMD in the 300 to 350 nm region produced HCl in $v'' = 0, 1,$ and 2. The population ratio at levels 0:1:2 is 1:0.48:0.31, corresponding to T = 600 K. The results are in sharp contrast with the UV flash photolysis of vinyl chloride itself, in which HCl is formed in a highly nonstatistical population.

Figuera and Tobar[27] have measured the quantum yield of disappearance of CMD at 313, 334, and 343 nm excitation. The yield is 0.9 ± 0.1 and is independent of pressure from a few torr to 1 atm of CMD. The quantum yield is also invariant with the addition of oxygen, propane, and *cis-*butene-2. The results suggest that the lifetime of the excited state must be 10^{-10} sec or less and that radicals or triplet states are not involved in decomposition.

Emission spectra were observed by low energy electron impact of CMD.[28] The spectra were weak $N_2(C^3\Pi \rightarrow B^3\Pi)$ bands and strong $CN(B^2\Sigma^+ \rightarrow X^2\Sigma^+)$ and $CH(A^2\Delta \rightarrow X^2\Pi)$ bands. The threshold energy of electrons for the production of $N_2(C^3\Pi)$ is 12.5 eV. Liu et al.[28] suggest that this energy corresponds to the formation of $CH_3ClC:$ and $N_2(C^3\Pi)$. The threshold energies for $CH(^2\Delta)$ and $CH(B^2\Sigma)$ production, as well as their companion fragments, were uncertain. Hence useful information on the heat of formation of CMD could not be obtained.

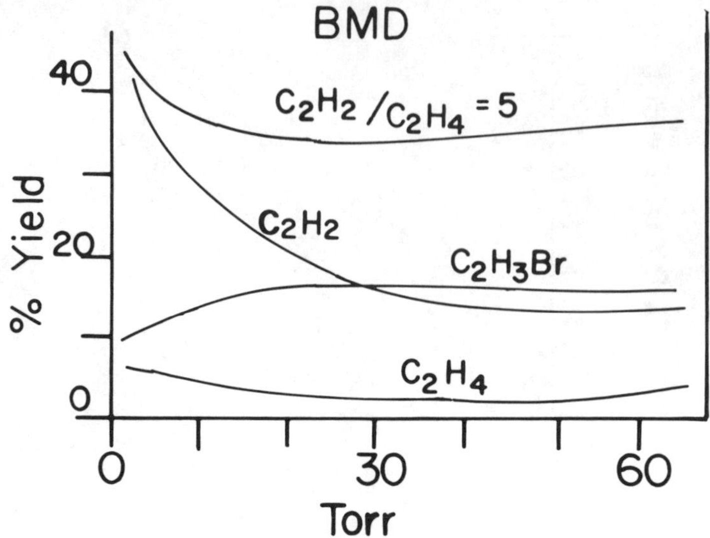

FIGURE 6.   Photolysis of BMD at 354 nm. Percentage of products are plotted against initial pressure of BMD. (From Crespo, M. T. et al., *J. Phys. Chem.*, 88, 5790. Copyright 1984 American Chemical Society.)

E. 3-Bromo-3-Methyldiazirine (BMD)

3-Bromo-3-methyldiazirine belongs to C_s symmetry. The UV spectrum[21] of BMD is similar to that of CMD shown in Figure 4, that is, the spectrum consists of two strong progressions of 1412 cm^{-1} and 802 cm^{-1} intervals. The former is the v_3' symmetric NN stretch, whereas the latter is the v_8' symmetric CC stretching frequency. The O–O band is at 353.8 nm (28257 cm^{-1}), slightly red to that of CMD. Results of the electron impact study are given in Tables 1 and 2. From the appearance threshold of $C_2H_3^+$ ion, the heat of formation of 72 kcal/mol is estimated.[32]

The photolysis products at 354 nm[29] are acetylene, ethylene, vinyl bromide, bromoethanes, and nitrogen. The quantum yield of BMD disappearance is 0.9 ± 0.15. The percentage of the products as a function of BMD pressure is presented in Figure 6. The ratio of acetylene to ethylene is about 5 and almost independent of the BMD pressure. The pressure dependence of the products can be explained by

$$BMD \xrightarrow{h\nu} CH_3BrC: + N_2 \tag{18}$$

$$CH_3BrC: \rightarrow C_2H_3Br^\dagger \tag{19}$$

$$C_2H_3Br^\dagger \rightarrow C_2H_2 + HBr \tag{20}$$

$$C_2H_3Br^\dagger \rightarrow C_2H_3 + Br \tag{21}$$

$$C_2H_3 + BMD \rightarrow C_2H_4 + R \tag{22}$$

where $C_2H_3Br^+$ signifies a vibrationally excited ground state C_2H_3Br molecule. The ratio C_2H_2/C_2H_4 of 5 indicates that the branching ratio, k_{20}/k_{21}, of $C_2H_3Br^\dagger$ decomposition is also 5, independent of the total pressure. Unimolecular theory suggests that $C_2H_3Br^\dagger$ containing about 80 kcal/mol dissociates into products with the C_2H_2/C_2H_4 ratio of 5. The 80 kcal/mol

Table 3
HEATS OF FORMATION OF DIAZIRINES

Diazirine[a]	Method
79	Electron impact;[3] from the appearance potential of CH_2^+ assuming CH_2^+ in the ground state
101	From appearance potential of CH_2^+ assuming CH_2^+ in the electronically excited state[13]
60.6[b]	Photon impact;[4] from the appearance potential of CH(A → X) fluorescence
< 66	No emission at 147 nm excitation[4]
CMD[a]	
55.6 ± 9.5	Thermal conductivity[6]
58	Electron impact;[32] from A.P.[c] of $C_2H_3^+$
>48	From A.P. of $C_2H_3Cl^+$, see Table 1
64	Estimate by Jones et al.[22] from the heat of formation of diazirine
51[b]	Estimated by Cadman et al.[20] and Frey et al.[23] from the heat of formation of diazirine
BMD[a]	
60[b]	Estimated by Cadman et al.[20] from ΔH_f° of diazirine
72	Electron impact;[32] from A.P. of $C_2H_3^+$;
>55	From A.P. of $C_2H_3Br^+$; see Table 1
DMD[a]	
64	Estimate by Jones et al.[22] from the heat of formation of diazirine

[a] kcal/mol.
[b] Recommended heats of formation.
[c] Appearance potential.

internal energy corresponds to 66% of the total available energy, that is, incident photon energy of 81 kcal/mol plus the exothermicity of 41 kcal/mol for BMD → C_2H_3Br + N_2. The addition of O_2 almost completely suppresses the formation of ethylene, indicating a radical precursor for ethylene production, as in Reaction 22.

Crespo et al.[29] suggest that the formation of 1,1-dibromoethane from the reaction

$$CH_3BrC: + HBr \rightarrow CHBr_2CH_3 \tag{23}$$

may not be important, because the addition of SF_6 to BMD and HBr mixtures does not reduce the sum of acetylene, ethylene, and vinyl bromide. If reaction 23 is important, the added SF_6 would promote the formation of 1,1-dibromoethane, and thus the sum of acetylene, ethylene, and vinyl bromide would be decreased. They suggest that 1,1-dibromoethane may be formed by processes such as

$$C_2H_3Br + Br \rightarrow C_2H_3Br_2 \tag{24}$$

$$C_2H_3Br_2 + HR \rightarrow C_2H_4Br_2 + R \tag{25}$$

The rate of isomerization of CH_3BrC to C_2H_3Br is unknown. In analogy with CH_3ClC, CH_3BrC may add stereospecifically to the double bond in competition with the isomerization, although such addition reaction has not been proven.

The heats of formation of four diazirines are listed in Table 3. Also, the heats of formation of various species used to derive the ΔH_f° values for diazirines are listed in Table 4.

Table 4
THE HEATS OF
FORMATION OF
SPECIES USED TO
CALCULATE THE
HEATS OF FORMATION
OF DIAZIRINES IN
TABLE 3

Species	ΔH_f° (kcal/mol)
HCl	-22.0[34]
C_2H_2	54.19[33]
C_2H_3Cl	8.4[33]
C_2H_3Br	18.7[33]
Cl	28.52 ± 2[34]
Br	28.2[34]
CH_2	92.25[5]
CH_2^+	332[12]
$C_2H_3^+$	269[12]
$C_2H_3Cl^+$	241[12]
$C_2H_3Br^+$	248[12]

F. Difluorodiazirine (DFD)

DFD, like diazirine, has C_{2v} symmetry. The absorption spectrum[30] of DFD resembles that of diazirine shown in Figure 1, but shifts toward longer wavelengths. For information, see Figure 7. The O–O band is at 352.3 nm.[31] The spectrum is composed of many sharp regularly spaced peaks between 282 and 251.5 nm. The vibrational analysis indicates that the spectrum consists of the v_1' symmetric NN stretching (1436 cm^{-1}) and v_4' symmetric CF_2 deformation (516 cm^{-1}) frequencies. DFD is chemically less reactive than either diazomethane or diazirine. It is stable in strong acids, but tends to explode. Thermal decomposition occurs above about 160°C. The heat of formation apparently has not been obtained. The mass spectra of DFD are included in Table 2.

The UV photolysis of pure DFD produces C_2F_4 and N_2 in analogy with the photolysis of diazirine.[14]

$$CF_2N_2 \xrightarrow{\ h\nu\ } CF_2 + N_2 \tag{26}$$

$$CF_2 + CF_2N_2 \rightarrow C_2F_4 + N_2 \tag{27}$$

The CF_2 radical adds stereospecifically to olefins to form difluorocyclopropanes, suggesting that the CF_2 radical is in a singlet state.[30] It also reacts with Cl_2, I_2, N_2O_4, and $ClNO_2$ to form corresponding difluoromethane derivatives.[30]

$$CF_2 + XY \rightarrow CF_2XY$$

$$X=Y=Cl$$

$$X=Y=I$$

$$X=Y=NO_2$$

$$X=Cl, Y=NO_2 \tag{28}$$

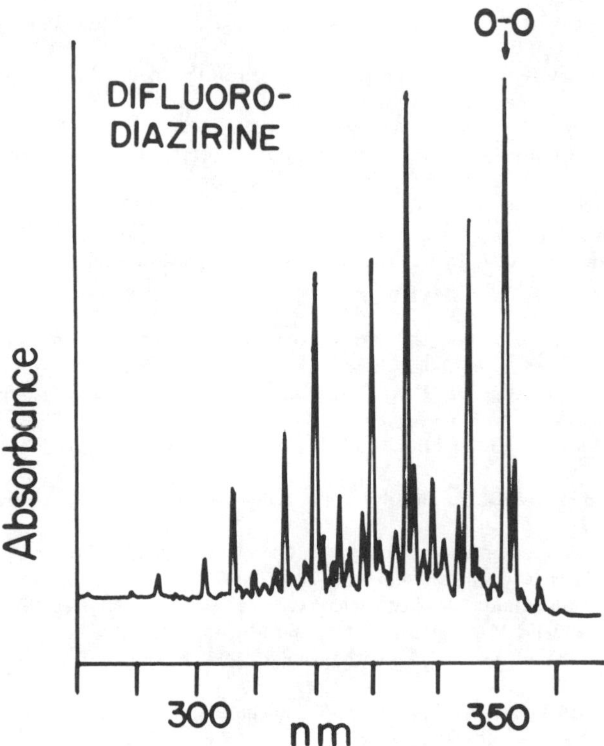

FIGURE 7.   The UV absorption spectrum of CF_2N_2. It is characterized by many sharp peaks between 282.0 and 351.5 nm; the molar absorption coefficient at 351.5 nm is 646.9 dm^3/mol/cm. (From Mitsch,, R. A., *J. Heterocyc. Chem.*, 159, 1964. With permission.)

REFERENCES

1. **Schmitz, E. and Ohme, R.,** Cyclo-diazomethane, *Tetrahedron Lett.*, 17, 612, 1961.
2. **Graham, W. H.,** A direct method of preparation of diazirine, *J. Am. Chem. Soc.*, 84, 1063, 1962.
3. **Paulett, G. S. and Ettinger, R.,** Mass spectra and appearance potentials of diazirine and diazomethane, *J. Chem. Phys.*, 39, 825, 1963.
4. **Laufer, A. H. and Okabe, H.,** Determination of the heat of formation of diazirine by photon impact, *J. Phys. Chem.*, 76, 3504, 1972.
5. **Okabe, H.,** *Photochemistry of Small Molecules,* John Wiley & Sons, New York, 1978, 258.
6. **Archer, W. H. and Tyler, B. J.,** Self-heating in the decomposition of 3-methyl-3-chlorodiazirine; determination of reaction exothermicity and correction of Arrhenius parameters, *J. Chem. Soc. Faraday Trans. I,* 72, 1448, 1976.
7. **Frey, H. M.,** The photolysis of the diazirines, in *Advances in Photochemistry,* Vol. 4, Noyes, W. A., Jr., Hammond, G. S., and Pitts, J. N., Eds., John Wiley & Sons, New York, 1966, 225.
8. **Liu, M. T. H.,** The thermolysis and photolysis of diazirines, *Chem. Soc. Rev.*, 11, 127, 1982.
9. **Vogt, J., Winnewisser, M., and Christiansen, J. J.,** The rotational-a-type spectra of ^{15}N-labeled diazirine, H_2CN_2, *J. Mol. Spectrosc.*, 103, 95, 1984; **Ettinger, R.,** Infrared spectrum of diazirine, *J. Chem. Phys.*, 40, 1693, 1964.
10. **Robertson, L. C. and Merritt, J. A.,** The electronic absorption spectra of n-diazirine, N_1^{15}-diazirine, d_1-diazirine, and d_2-diazirine, *J. Mol. Spectrosc.*, 19, 372, 1966.
11. **Merritt, J. A.,** The ultraviolet spectrum of diazirine, *Can. J. Phys.*, 40, 1683, 1962.
12. **Rosenstock, H. M., Draxl, K., Steiner, B. W., and Herron, J. T.,** Energetics of gaseous ions, *J. Phys. Chem. Ref. Data,* Suppl. 1, 6, 1977.

13. **Bell, J. A.,** On mass spectra and appearance potentials of diazirine and diazomethane, *J. Chem. Phys.,* 41, 2556, 1964.
14. **Frey, H. M. and Stevens, I. D. R.,** The formation of methylene by the photolysis of diazirine (cyclodiazomethane), *Proc. Chem. Soc.,* 79, 1962.
15. **Amrich, M. J. and Bell, J. A.,** Photoisomerization of diazirine, *J. Am. Chem. Soc.,* 86, 292, 1964.
16. **Moore, C. B. and Pimentel, G. C.,** Matrix reaction of methylene with nitrogen to form diazomethane, *J. Chem. Phys.,* 41, 3504, 1964.
17. **Frey, H. M. and Stevens, I. D. R.,** The photolysis of 3-methyl-diazirine (cyclodiazoethane), *J. Chem. Soc.,* 1700, 1965.
18. **Frey, H. M. and Stevens, I. D. R.,** The photolysis of dimethyl-diazirine, *J. Chem. Soc.,* 3514, 1963.
19. **Robertson, L. C. and Merritt, J. A.,** Electronic absorption spectra of 3,3'-dimethyldiazirine, *J. Chem. Phys.,* 56, 2919, 1972.
20. **Cadman, P., Engelbrecht, W. J., Lotz, S., and Van der Merwe, S. W. J.,** Energy distribution in reaction products; photolysis of 3-halo-3-methyldiazirines, *J. S. Afr. Chem. Inst.,* 27, 149, 1974.
21. **Robertson, L. C. and Merritt, J. A.,** The electronic absorption spectra of bromomethyldiazirine and chloromethyldiazirine, *J. Mol. Spectrosc.,* 24, 44, 1967.
22. **Jones, W. E., Wasson, J. S., and Liu, M. T. H.,** Photolysis of 3-chloro-3-methyldiazirine, *J. Photochem.,* 5, 311, 1976.
23. **Frey, H. M. and Penny, D. E.,** Photolysis of 3-methyl-3-chlorodiazirine, *J. Chem. Soc. Faraday Trans. I,* 73, 2010, 1977.
24. **Figuera, J. M., Perez, J. M., and Tobar, A.,** Energy distribution and mechanism in 3-chloro-3-methyldiazirine photolysis, *J. Chem. Soc. Faraday Trans. I,* 74, 809, 1978.
25. **Moss, R. A. and Mamantov, A.,** Methylchlorocarbene, *J. Am. Chem. Soc.,* 92, 6951, 1970.
26. **Becerra, R., Castillejo, M., Figuera, J. M., and Martin, M.,** HCl vibrational distribution in the dissociation of vinyl chloride activated by flash photolysis of 3-chloro-3-methyldiazirine, *Chem. Phys. Lett.,* 100, 340, 1983.
27. **Figuera, J. M. and Tobar, A.,** Photodecomposition quantum yields of 3-chloro-3-methyldiazirine in the gas phase, *J. Photochem.,* 10, 473, 1979.
28. **Liu, M. T. H., Tanaka, T, Hirotsu, T., Fukui, K., Fujita, I., and Kuwata, K.,** Excitation and dissociation of 3-chloro-3-methyldiazirine and 1-pyrazoline by low energy electron impact, *J. Phys. Chem.,* 84, 3184, 1980.
29. **Crespo, M. T., Figuera, J. M., Rodriguez, J. C., and Utrilla, R. M.,** Photolysis of 3-bromo-3-methyldiazirine, *J. Phys. Chem.,* 88, 5790, 1984.
30. **Mitsch, R. A.,** Difluorodiazirine. I. Physical and spectral properties, *J. Heterocyc. Chem.,* 1, 59, 1964; II. Difluoromethane derivatives (1), *J. Heterocyc. Chem.,* 1, 233, 1964; III. Synthesis of difluorocyclopropanes, *J. Am. Chem. Soc.,* 87, 758, 1965.
31. **Simmons, J. D., Bartky, I. R., and Bass, A. M.,** Vibrational fundamentals of CF_2N_2 from the ultraviolet absorption spectrum, *J. Mol. Spectrosc.,* 17, 48, 1965; **Bjork, C. W., Craig, N. C., Mitsch, R. A., and Overend, J.,** Infrared and Raman spectra of CF_2N_2; evidence for a diazirine structure, *J. Am. Chem. Soc.,* 87, 1186, 1965.
32. **Engelbrecht, W. J. and Loubser, G. J.,** Determination of the heats of formation of diazirines, *J. S. Afr. Chem. Inst.,* 28, 191, 1975.
33. **Stull, D. R., Westrum, E. F., Jr., and Sinke, G. C.,** *The Chemical Thermodynamics of Organic Compounds,* John Wiley & Sons, New York, 1969.
34. **Stull, D. R. and Prophet, H., Eds.,** JANAF Thermochemical Tables, 2nd ed., National Standard Reference Data Series, U.S. National Bureau of Standards, Washington, D.C., 1971, 37.

Chapter 8

THE SELECTIVITY OF CARBENES GENERATED FROM DIAZIRINES

Michael P. Doyle

TABLE OF CONTENTS

I. INTRODUCTION

Diazirines have become an important source for carbene generation. Their synthesis affords entry into classes of carbenes whose production often cannot be achieved by alternative methodologies. In contrast to diazo compounds, which are characteristically sensitive to acid-promoted decomposition, diazirines are stable toward acids. Consequently, diazirines complement diazo compounds as carbene sources and the versatility of diazirines for carbene generation depends primarily on methods for their synthesis.

This chapter focuses on the selectivity of carbenes generated from diazirines. Consequently, it will be concerned with carbenes and not with the pathway taken in their generation. The questions of diazirine-diazoalkane interconversion[1] and the mechanism for diazirine decomposition to carbenes[2] are addressed elsewhere in this book. Emphasis is placed here on those carbenes that have been produced from diazirines, alternative methods for their generation, and the forms of selectivity that are evidenced in their transformations, especially reactivity, chemoselectivity, regioselectivity, and stereoselectivity.

"Selectivity of carbenes" is a broad topic addressed in a variety of books and reviews. Foremost among these have been the serial publication *Reactive Intermediates* edited by Jones and Moss[3-5] and their earlier monographs entitled *Carbenes*.[6,7] Other texts, especially Kirmse's *Carbene Chemistry,* have also provided limited treatment of carbene selectivity,[8-10] and several reviews have described specific developments related to this topic.[11-14] Although overlap cannot be avoided, this chapter demonstrates the unique advantages of diazirines for carbene generation and describes the specific reaction characteristics of carbenes produced from diazirines.

II. DIAZIRINES AS CARBENE SOURCES

A. Comparative Methods for Carbene Generation: "Free" Carbenes

Only a limited number of transformations afford entry into general classes of carbenes; both diazirines and diazo compounds offer the greatest versatility. Photolytic or thermal decomposition of diazirines or diazo compounds results in the extrusion of dinitrogen and production of the "free" carbene (Equation 1).

$$R_2C \overset{N}{\underset{N}{\parallel}} \longrightarrow R_2C: + N_2 \qquad (1)$$

$$R_2C{=}N_2$$

In contrast, base-induced α-elimination as an alternative method for carbene generation (Equation 2)

$$R_2C\overset{H}{\underset{X}{\big\langle}} + M^+Base^- \underset{+HBase}{\overset{-HBase}{\rightleftharpoons}} R_2C\overset{M}{\underset{X}{\big\langle}} \longrightarrow \begin{array}{c} \text{carbenoid} \\[1ex] \downarrow{-MX} \\[1ex] R_2C: \end{array} \qquad (2)$$

Table 1
RELATIVE REACTIVITIES OF ALKENES TOWARD
PHENYLBROMOCARBENE[15]

Alkene	PhCBrN$_2$[a] hv, 25°C	PhCHBr$_2$, *t*-BuOK/*t*-BuOH, 25°C	
		With 18-crown-6	Without 18-crown-6
Me$_2$C=CMe$_2$	4.4	4.1	1.6
Me$_2$C=CHMe	2.5	2.4	1.3
Me$_2$C=CH$_2$	1.00	1.00	1.00
c-MeCH=CHMe	0.53	0.50	0.29
t-MeCH=CHMe	0.25	0.24	0.15

[a]   Phenylbromodiazirine.

often produces a **carbenoid species** *that, although undergoing reactions which are characteristic of free divalent carbon species, exhibits reactivities and/or selectivities that differ from those of carbenes generated from diazirines or diazo compounds.*

The solution chemistry of carbenes has required a standard for evaluation of the "freeness" of these highly reactive species, and diazirine decomposition has achieved that status. Diazirines are normally stable under ordinary conditions, they are amenable to purification by chromatography, and they undergo either photochemical or thermal decomposition with extrusion of dinitrogen. In what has become a classical experimental probe for "free" carbenes, Moss and Pilkiewicz[15] discovered that phenylhalocarbenes produced by base-induced α-elimination in the presence of 18-crown-6 (path A) have the same relative reactivities in olefin addition reactions (Scheme 1) as do the same carbenes generated by photolysis (path B) of phenylhalodiazirines (Table 1). In the absence of 18-crown-6, relative reactivities for cyclopropanation by base-induced α-elimination (path C) are appreciably lower,[16] which suggests that the reactive intermediates are probably carbene complexes with either KX or KO-*t*-Bu. The discovery of the influence of 18-crown-6 on carbene selectivities not only reinforces the proposition that photolysis of diazirines generates free carbenes but also affords a useful methodology to distinguish "free" carbenes from carbenoids in cases where neither the diazirine nor the diazo precursor to the carbene is available.[17-19]

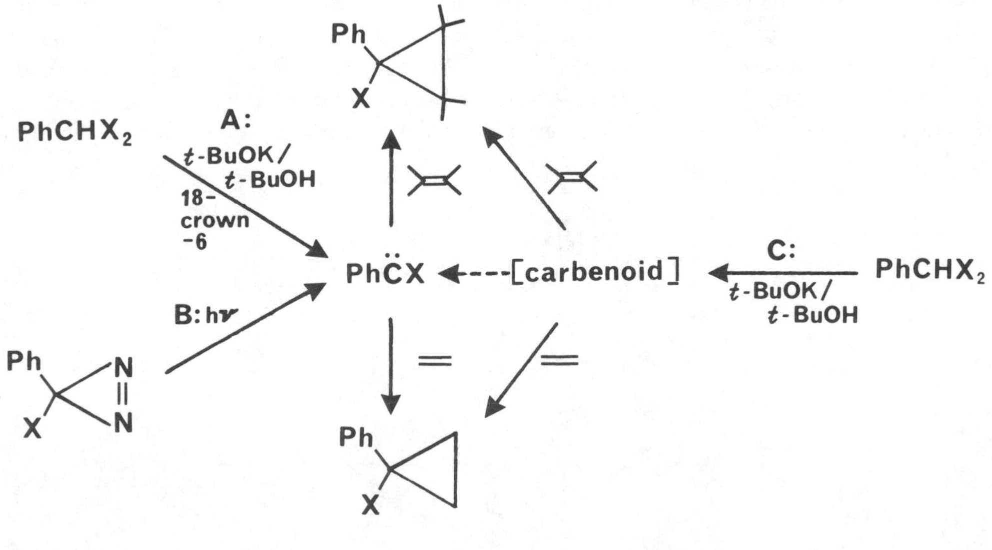

SCHEME 1.

Table 2

PRODUCT DISTRIBUTION IN INTRAMOLECULAR REACTIONS OF ETHYLMETHYLCARBENE

Relative yield C_5H_{10} isomers

Method (reference)						
Br/Br — MeLi (19)	1	74	25	–	–
Br/Cl — MeLi (19)	2	74	23	–	–
=NNHTs — NaOMe (20)	5	67	28	0.5	0
N=N — Δ (21)	3	67	30	0.5	–
N=N — hν (21)	23	38	35	3.7	0.3

Intramolecular insertion reactions and 1,2-hydrogen migrations of carbenes also exhibit selectivities that are dependent on the method for carbene generation. The characteristics of this dependence are exemplified in results from various methods employed to produce ethylmethylcarbene (Table 2). The product distribution obtained from transmetallation-elimination[20] clearly differs from that derived from either 2-diazobutane[21] or thermal diazirine decomposition[22] in the ratio of 2-butene to 1-butene, in the ratio of *trans*- to *cis*-2-butene, and in the production of 1-methylcyclopropane. More important, however, is the discovery that thermal and photolytic decomposition of 3-ethyl-3-methyldiazirine exhibit substantial differences in selectivity. Analogous differences in product distribution from thermal and photolytic decomposition of **1** (Equation 3) have been reported,[23] and the composite results have been explained[23,24] as due to generation of a more energetic intermediate in the photolytic decomposition of diazirines. Similar observations have been made for thermal and photolytic decomposition of diazoalkanes.[25]

$$(3)$$

At first glance, the differences between thermal and photolytic diazirine decomposition in product distributions from intramolecular reactions appear to be in conflict with conclusions drawn from results with phenylhalocarbenes in intermolecular cyclopropanation reactions.[15] However, neither the carbene species nor the transformations being compared are similar. The photolytically generated phenylhalocarbenes are clearly not vibrationally excited species in their reactions with alkenes. Whether this is a function of the carbene or the transformation remains to be determined.

B. Comparative Methods for Carbene Generation: Competing Reactions

Diazirines and diazo compounds are remarkably different in their sensitivity toward acids. Diazo compounds, particularly those that are aliphatic, are especially sensitive to acid-promoted decomposition,[26,27] and this sensitivity has often dominated consideration of a large class of diazo compounds for synthesis and carbene generation. In contrast, diazirines exhibit remarkable chemical stability,[28] which has permitted studies of carbene generation and reactions under a variety of conditions not readily accessible with other precursors.

Thermal and photochemical decomposition of diazo compounds or diazirines results in the formation of azines. Carbene dimers are also common by-products in these reactions. Azine formation has been viewed as a bimolecular reaction between two diazo compounds followed by elimination of dinitrogen (path A, Scheme 2),[29,30] as a coupling between a

SCHEME 2.

carbene and a diazo compound (path B),[31] and as an addition of a carbene to a diazirine followed by rearrangement (path C).[32,33] However, the latter two mechanisms do not explain results obtained by Liu and Ramakrishnan[34] who showed that thermal decomposition of either phenylchlorodiazirine or phenylmethyldiazirine resulted in nearly quantitative azine formation, but that decomposition of an equimolar mixture of these diazirines under identical conditions produced the mixed carbene dimer without azine. Formation of the mixed carbene dimer was reported to result from nucleophilic attack of the diazo compound, presumed to have been formed from the diazirine, on the electron deficient carbene (Equation 4). Loss of dinitrogen from the resulting zwitterion (**1**) completed formation of the observed alkene products (1:1 isomer ratio).

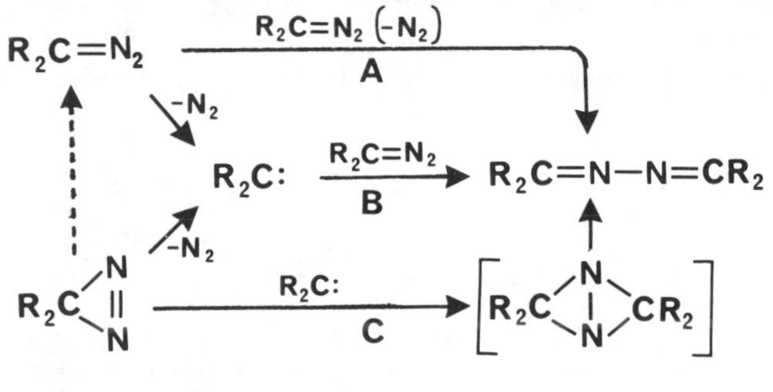

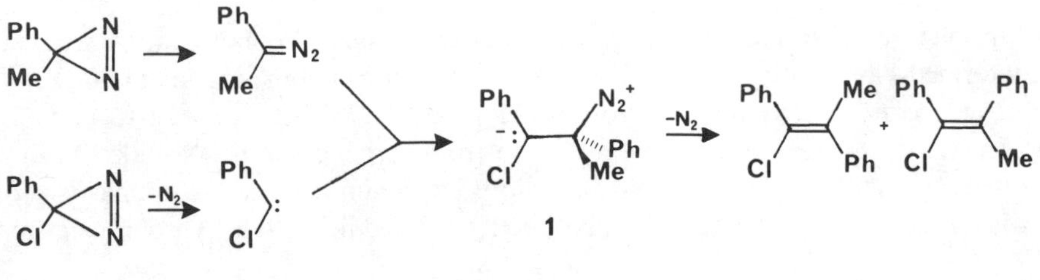

(4)

Alkene formation through thermal decomposition of diazirines in the presence of diazo compounds appears to be a general transformation (Equation 5).[35] Diazocarbonyl compounds are stable under conditions that cause dinitrogen extrusion from phenylchlorodiazirine, and mixed azines are not by-products of these reactions. Product isomer ratios appear to be governed by steric and stereoelectronic factors.

$$Z = OEt \quad\quad 64 \quad 36$$
$$N(i\text{-}Pr)_2 \quad\quad 77 \quad 23$$
$$Ph \quad\quad\quad 81 \quad 19 \tag{5}$$

If alkenes are formed by electrophilic carbene addition to diazo compounds, then the same pathway is unlikely to account for azine formation (path B of Scheme 2). Although path A is consistent with the limited available data, assuming that diazirine isomerization is the rate-determining step, it is equally likely that carbene association with the diazirine promotes subsequent rearrangement to azine (Equation 6). According to this scheme, alkene formation is dependent on the extent of diazirine isomerization, whereas azine formation results from the unrearranged diazirine. Further investigations are warranted.

$$(6)$$

If the diazirine undergoes isomerization to a diazo compound, dipolar addition of the diazo compound to alkenes may be expected (Equation 7).[36,37] Diazoalkanes are more reactive

$$(7)$$

toward dipolar addition than are aryldiazomethanes or diazocarbonyl compounds, but their initially formed 1-pyrazolines are more stable toward either tautomerization to 2-pyrazolines[38] or dinitrogen extrusion to form cyclopropane products.[39] This pathway to cyclopropane formation from diazirines can be distinguished from carbene addition to alkenes in certain cases by (1) observation of 2-pyrazolines, whose formation from 1-pyrazolines is promoted by Lewis bases,[38] (2) the absence of stereospecificity in addition to the carbon-carbon double bond, and (3) production of formal vinyl C–H insertion products.[39] Thermal decomposition of diazirines should favor dipolar addition so long as the rates for diazirine isomerization and dipolar addition are greater than that for carbene formation. However, cyclopropane formation via isomerized diazirines is yet to be reported.

Diphenyldiazomethane undergoes dipolar cycloaddition to substituted styrenes at 50°C.[40] Loss of dinitrogen to form cyclopropane products (70 to 80% yields) occurs in a subsequent fast step. In contrast, photochemical decomposition of diphenyldiazomethane at or below 0°C also results in cyclopropane formation but, under these conditions, via triplet diphenylcarbene. In this investigation the dipolar addition pathway was identified from its unique kinetic characteristics. The preparation and thermal decomposition of 3,3-diphenyldiazirine should afford complementary information, especially if isomerization of diphenyldiazomethane fails to occur.

III. CARBENIC SELECTIVITY IN CYCLOPROPANATION REACTIONS

A. Relative Reactivity

As carbene sources, particularly for heteroatom stabilized carbenes,[41,42] diazirines have played an essential role in the development of quantitative correlations of carbene reactivities toward alkenes. Diazirines have provided a single generative source for carbenes that could be produced at the same temperature and in the absence of reactants capable of association with these reactive intermediates, as in the base-promoted methodologies. Since this development is discussed in detail elsewhere,[3-5,14] only the basic strategy and general features of these correlations will be presented here. A renaissance in understanding carbene selectivity has resulted from these investigations.

Following from earlier relative reactivity studies of Skell et al.[43-46] and Doering et al.,[47,48] Moss et al.[49,50] have defined a ''carbene selectivity index'', m_{CXY}, from the equation relating relative reactivities for the carbene in question, :CXY, to relative reactivities for dichlorocarbene reactions with the same set of alkenes:[49,50]

$$\log(k_i/k_{\text{isobutene}})_{CXY} = m_{CXY}\log(k_i/k_{\text{isobutene}})_{CCl_2}^{25°C} + c \tag{8}$$

Five alkenes, tetramethylethylene, trimethylethylene, isobutene, and *cis*- and *trans*-2-butene, were chosen for the reference set, and isobutylene was selected as the standard. Dichlorocarbene was chosen to be the reference carbene with $m_{CCl_2} = 1.00$ at 25°C. Only singlet carbenes were chosen for investigation, and carbenoid species were excluded whenever this characteristic could be identified. Representative data for carbenes produced by Moss through photochemical decomposition of diazirines in the presence of this standard set of alkenes, relative to :CCl$_2$ are given in Table 3.

An initial set of six carbenes was subjected to multiple linear regression analysis of the dependence of m_{CXY} on σ_{R+} and σ_I constants:[55]

$$m_{CXY} = c_R\Sigma_{X,Y}\sigma_{R+} + c_I\Sigma_{X,Y}\sigma_I + c_o \tag{9}$$

and $c\sigma_{R+}$, c_I, and c_o were calculated to be -0.94, $+0.69$, and -0.27, respectively.[50] With a more complete set of m_{CXY} values from ten carbenes, this equation was further refined:[49]

$$m_{CXY} = -1.10\ \Sigma_{X,Y}\sigma_{R+} + + 0.53\ \Sigma_{X,Y}\sigma_I - 10.31 \tag{10}$$

Graphical presentation of experimentally determined $m_{CXY}^{25°}$ values vs. those calculated from Equation 10 (Figure 1) describes a linear relationship (slope = 1.00) with a correlation coefficient of 0.971. Values of σ_{R+} afforded the best correlation in the three-parameter equation.[49]

Table 3
RELATIVE REACTIVITIES OF CARBENES PHOTOLYTICALLY GENERATED FROM DIAZIRINES

	$k_i/k_{isobutylene}$, 25°C					
Alkene	CCl$_2$[a]	PhCCl[b]	PhCBr[c]	MeCCl[d]	c-C$_3$H$_5$CCl[e]	BrCCOOEt[f]
Me$_2$C=CMe$_2$	8.9	5.1	4.4	3.9	2.4	1.9
Me$_2$C=CHMe	3.1	3.2	2.5	2.4	1.8	1.4
Me$_2$C=CH$_2$	1.00	1.00	1.00	1.00	1.00	1.00
c-MeCH=CHMe	0.27	0.37	0.53	0.74	0.67	0.70
t-MeCH=CHMe	0.18	0.20	0.25	0.52	0.46	0.62
m_{CXY}	1.00	0.83	0.70	0.50	0.41	0.29

[a] From KO-t-Bu-mediated α-elimination with CHCl$_3$, Reference 5, p. 221.
[b] Reference 51.
[c] Reference 52.
[d] Reference 53.
[e] Reference 54.
[f] Reference 49.

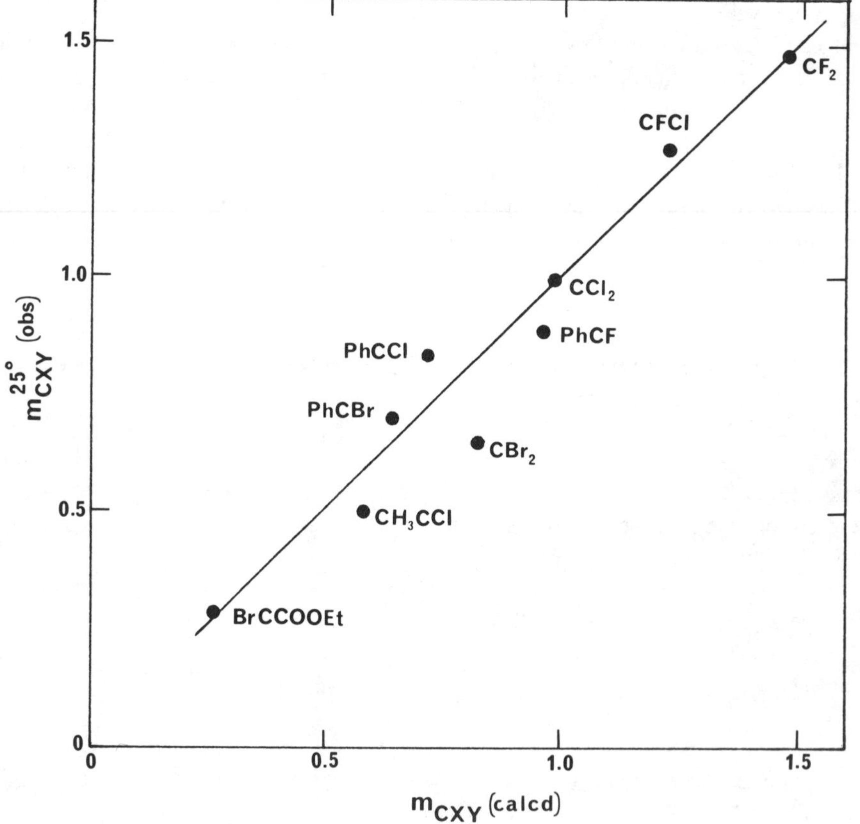

FIGURE 1. Correlation of experimentally determined carbene selectivity indices, m_{CXY}, with those calculated from Equation 10.

Table 4
**EXPERIMENTALLY DETERMINED CARBENE
SELECTIVITY INDICES**

CXY	Source	m_{CXY}^{obs}	m_{CXY}^{calcd}	m	Ref.
CF_2	c-N_2CF_2,hν[a]	1.48	1.47	0.01	49
CFCl	$CHFCl_2$,t-BuOK	1.28	1.22	0.06	49
CCl_2	$CHCl_3$,t-BuOK	1.00	0.97	0.03	5
MeSCCl	$MeSCHCl_2$,t-BuOK	0.91	$(1.42)^b$	(-0.51)	56
PhCF	$PhCHFBr$,t-BuOK	0.89	0.96	-0.07	57
PhCCl	c-$N_2C(Cl)Ph$, hν	0.83	0.71	0.12	51
PhCBr	c-$N_2C(Br)Ph$,hν	0.70	0.64	0.06	52
CBr_2	$CHBr_3$,t-BuOK	0.65	0.82	-0.17	19
MeCCl	c-$N_2C(Cl)Me$,hν	0.50	0.58	-0.08	53
CpCCl	c-$N_2C(Cl)Cp$,hν	0.41	$(0.73)^c$	(-0.32)	54
BrCCOOEt	$N_2CBrCOOEt$,hν	0.29	0.26	0.03	49
CF_3CCl	c-$N_2C(Cl)CF_3$,hν	0.19	0.31	-0.12	58

[a] c-N_2 denotes diazirine.

[b]   Value of σ_{R+} used for this calculation is questionable; see Reference 49.

[c]   Value of σ_{R+} is defined for a "bisected" rather than a "twisted" p-cyclopropyl-
tert-cumyl cation.

The approach defined by Moss and co-workers has allowed carbene reactivities to be evaluated on the same scale (Equation 10) for the first time. Substantial deviation from this correlation, as is seen in Figure 1 for CBr_2, suggests the operation of factors that are not incorporated in the σ_{R+} and σ_I values. Experimentally determined values of m_{CXY} are given in Table 4 along with those predicted from Equation 10. The numerical difference between observed and calculated m values, Δm, indicates their correspondence. Calculated values for m_{CXY} can be obtained when reliable σ values are available.

Recognizing an apparent relationship between alkene ionization potentials (IP) and σ_I values,[64] Freeman surveyed carbene reactivities and found reasonably good linear relationships between log $(k_i/k_{isobutylene})$ and IP values of as many as seven alkenes for several carbenes.[65] The slopes of these lines generally follow the m_{CXY} values obtained by Moss, although data points for *cis*- and *trans*-2-butene were often not included for slope determinations because these points fell off the line in four of the five cases reported. An earlier report by Skell defined a similar correlation for CCl_2.[46] Investigation of these correlations with a larger alkene data base has recently been reported.[66] Using a set of 11 olefins ranging from isobutylene to tricyclopropylethylene, Schoeller and co-workers showed that the relative reactivity of CCl_2 increases as the π-ionization potential of the olefin decreases. However, a direct linear relationship with IP was impaired by a large scattering of points. As a consequence, a two-parameter equation,

$$\ln k_{rel} = a(IP)_{olefin} + b(HOMO)_{olefin} + c \qquad (11)$$

where $(HOMO)_{olefin}$ is the larger atomic orbital coefficient in the HOMOs of the various olefins, was proposed, and correlation was improved (r^2 between 0.96 and 0.84). Values obtained for relative reactivities of carbenes ($S_{CXY} = a_{CXY}/a_{CCl2}$) agree remarkably well with those obtained by Moss (Table 5). Whereas Equation 10 offers predictability for carbene selectivity based on a standard set of alkenes, Equation 11 can be used to predict the relative reactivity of a specific olefin so long as the values of a, the π-ionization potential of the olefin, and the orbital coefficients of its HOMOs are known. The two equations are a powerful complement for the prediction of carbene reactivities with alkenes.

Table 5
CARBENE SELECTIVITY INDICES

Carbene	a_{CXY}[a]	S_{CXY}[b]	m_{CXY}[c]	PI[d]
CF_2	−8.75	1.51	1.48	0.85
CFCl	−7.41	1.28	1.28	0.61
CCl_2	−5.80	1.00	1.00	0.26
MeSCCl	−5.51	0.95	0.91	0.70
PhCF	−5.09	0.88	0.89	0.83
PhCCl	−4.67	0.81	0.83	
PhCBr	−3.88	0.67	0.70	
CBr_2	−3.70	0.64	0.65	
MeCCl	−2.82	0.49	0.50	0.44
CpCCl	−2.30	0.40	0.41	
BrCCOOEt	−1.66	0.29	0.29	

[a]   From Equation 11 (Reference 65).
[b] $S_{CXY} = a_{CXY}/a_{CCl_2}$.
[c]   From Table 4.
[d]   Reference 78.

Table 6
RATIO OF RELATIVE REACTIVITIES FOR CARBENE ADDITIONS TO ISOMERIC 2-BUTENES[a]

Carbene	k_{cis}/k_{trans}	m_{CXY}	Carbene	k_{cis}/k_{trans}	m_{CXY}
CF_2	0.88	1.48	CBr_2	1.26	0.65
CFCl	1.44	1.28	MeCCl	1.42	0.50
CCl_2	1.50	1.00	CpCCl	1.46	0.41
PhCCl	1.85	0.83	BrCCOOEt	1.13	0.29
PhCBr	2.12	0.70	CF_3CCl	1.42	0.19

[a]   Data taken from references listed for Table 4.

Substituted styrenes have been used in several cases in an alternative approach to evaluation of carbenic relative reactivities.[50,59-63] However, exact comparisons between carbenes have been limited by the diversity of reaction conditions employed for their determination and by the absence of uniform correlation with a single set of σ values. In one case, a diazo compound employed as a carbene precursor underwent competitive dipolar addition with substituted styrenes.[40]

Relative reactivities in carbene addition reactions with the standard set of alkenes (Table 3) are not apparently strongly influenced by steric effects. Modification of Equation 10 to include a steric term did not improve the correlation,[49] and the coefficients for the steric terms were order-of-magnitude smaller than those for the electronic terms. Also, if steric effects were important, one would have expected curvature in plots obtained from Equation 8. With few exceptions,[11,49,58] this has not been observed. However, comparison of relative reactivities for carbene addition to the isomeric *cis*- and *trans*-2-butenes suggests that steric factors may play a role in these reactions. The ratio k_{cis}/k_{trans} does not vary uniformly with m_{CXY} (Table 6). Instead, a maximum value is reached with PhCBr, which is also the bulkiest of the listed carbenes. Furthermore, whereas the *cis* olefin is more reactive than its *trans* isomer in electrophilic addition reactions involving three-membered cyclic activated complexes when steric effects are not important,[64] k_{cis}/k_{trans} is less than 1.0 for addition reactions of the electrophilic CF_2.

Table 7
**RELATIVE RATES FOR REACTIONS OF CCl_2
AND CBr_2 WITH ALKENES 3—5 AT 25°C[67]**

	CCl_2 k_{rel}			CBr_2 k_{rel}		
R	3	4	5	3	4	5
CH_3	320	1000	12.0	500	1000	
CH_2CH_3	210		9.1	480		68.0
$CH(CH_3)_2$	77	280	4.4	160	350	30.0
$CH(CH_2CH_3)_2$	29			60		
$C(CH_3)_3$	14	110	0.47	38	185	2.5

Giese et al. have recently challenged the minimal importance accorded steric effects in carbene addition reactions.[67] Reporting relative reactivities for reactions of CCl_2 and CBr_2 with 3[68] and 4[67] along with similar results obtained for **5** (Table 7) by Moss et al.[19] who find that the relative reactivities of CBr_2 and CCl_2 using the expanded set of alkenes (3—5) cannot be correlated with each other. Each set of alkenes gives its own linear free energy relationship. This outcome is not so surprising, because substantial structural changes such as those enacted on the reacting systems by **3** and **4** invite unique steric interactions. This approach is leading toward a common correlation of CX_2 relative reactivities with polar and steric substituent effects. Whether it can extend beyond dihalocarbenes remains to be determined.

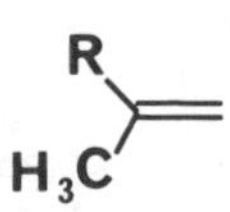

3 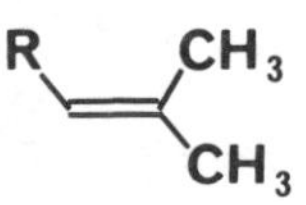   **4** 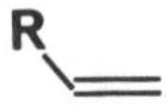   **5**

The influence of steric effects on carbene reactivities has been critically evaluated with trifluoromethylchlorocarbene in comparison with dichlorocarbene.[58] Among the carbenes included in Table 4, CF_3CCl is the least discriminating. Relative reactivities for addition to the standard set of alkenes employed for m_{CXY} determination range from 0.62 (*trans*-MeCH=CHMe) to 1.17 (Me$_2$C=CHMe) with Me$_2$C=CMe$_2$, ordinarily the most reactive olefin, exhibiting a relative reactivity of 0.92. The data indicate that CF_3CCl experiences less steric hindrance than CCl_2 toward addition, and this conclusion has been reinforced by correlating $\log k_{rel}$ with Taft's E_s values[69] using 1-butene, 3-methyl-1-butene, and 3,3-dimethyl-1-butene as substrates. Values of δ (0.41 for CF_3CCl and 0.88 for CCl_2) show that CF_3CCl is less susceptible to steric hindrance during its addition reactions. Relative to CH_3CCl, the CF_3 group strongly destabilizes CF_3CCl. As a result, reaction occurs via an "early" transition state which is less sensitive to the steric demand of carbene substituents than is that of more selective, structurally smaller carbenes such as CCl_2. The unexpected low reactivity of CF_3CCl toward Me$_2$C=CMe$_2$ may be a consequence of an approach of this reaction rate to diffusion control,[58] but other factors cannot be ruled out.

The difference between observed and calculated *m* values for cyclopropylchlorocarbene (Table 4) is significant, and resolution of this discrepancy offered insight into its conformational stabilities and reactivities of its limiting conformers.[54] *A priori*, cyclopropylchlorocarbene is expected to prefer the bisected conformation **6b** in which the "bent" cyclopropyl σ bonds can favorably interact with the vacant carbenic p orbital.[70,71]

Table 8
TEMPERATURE
DEPENDENCE OF m_{CXY}[72]

Temp., C°	m_{CF_2}	m_{CCl_2}	m_{CBr_2}
25	1.40	1.00	0.62
65	1.19	1.00	0.91
125	0.83	1.00	1.17

6b   **6t**

Values for σ_{R+} and σ_I, obtained from solvolyses of *p*- and *m*-cycloproyl-*tert*-cumyl chlorides,[72] reflect the stabilization afforded by the bisected conformation. However, Moss anticipated that **6b** would encounter substantial steric hindrance in electrophilic addition to alkenes, so that addition should actually occur via a twisted conformation (**6t** in the limit). *Ab initio* calculations established the general relationship (Equation 12) between "carbene stabilization energies"

$$\Delta E_{stab} = 28.6 m_{CXY} + 12.8 \qquad (12)$$

and m_{CXY}, for 12 carbenes with only CCl_2 showing substantial deviation. Calculation of ΔE_{stab} for **6b** and **6t** afforded values of 35.9 and 26.4 kcal/mol, respectively. From Equation 12 the carbene selectivity indices $m_{6b}^{calcd} = 0.81$ and $m_{6t}^{calcd} = 0.48$ were generated. Most significantly, the experimental m_{CXY}^{obs} value of 0.41 corresponds to m_{6t}^{calcd}, whereas m_{CXY}^{calcd} obtained from Equation 10 corresponds to m_{6b}^{calcd}.

Although correlations according to Equations 8 and 10 are the general rule, some notable exceptions exist. As mentioned earlier, CF_3CCl is a borderline case. In addition, HCCOOEt,[49] which is even less discriminating than CF_3CCl, and cyclobutylchlorocarbene,[54] which exhibits reactivities that are similar to those of **6t**, are not correlateable with CCl_2 according to Equation 8. With HCCOOEt, as with CF_3CCl, diffusion effects may be responsible, although competition between addition and carbon-hydrogen insertion, particularly for the highly branched olefins, is also probable. With cyclobutylchlorocarbene, steric factors have been offered as an explanation.[54]

Giese et al.[67,73-76] have noted that carbene selectivities for CF_2, CCl_2, and CBr_2 are temperature dependent. Representative data are shown in Table 8, and complementary equations to that of Equation 10 which equate resonance and inductive σ constants with m_{CXY} values at 60°C ($c_R = -1.1$, $c_I = -2.6$, and $c_o = +2.8$) and at 120°C ($c_R = -1.0$, $c_I = -7.4$, and $c_o = +7.1$) have been obtained.[75] Increasing temperature has a negligible effect on the resonance term, but the term responsive to inductive effects shows a dramatic dependence on temperature. These observations have led Giese to propose that the initial interaction between a carbene and an olefin is that of a charge-transfer complex **7** (Equation 13).

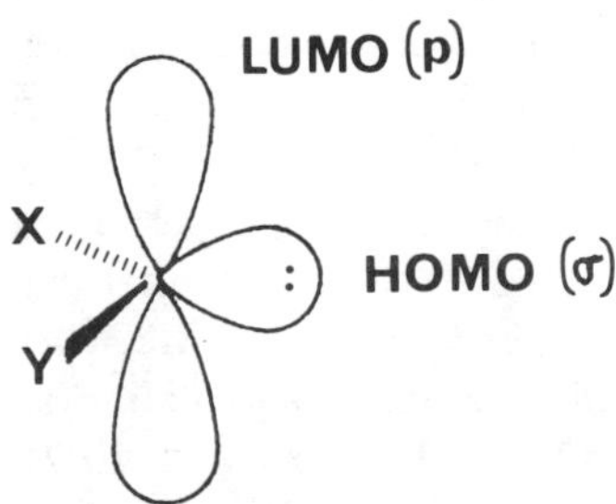

$$7 \qquad\qquad (13)$$

Skell had previously offered a similar conclusion.[43] This feature of carbene reactions is discussed in detail later in this chapter.

The temperature in a plot of log(relative reactivity) vs. 1/T(Kelvin) where carbenes exhibit the same selectivity is the *isoselective temperature*. This temperature, T_{iso}, is 90 ± 10°C for CX_2 reactions with methyl substituted alkenes[77] and also with **3** (R = i − Pr).[67] Below T_{iso}, the selectivity order obtained by Moss,[14] CF_2 > CFCl > CCl_2 > CBrCl > CBr_2, is observed; above T_{iso}, the order is inverted.[74] Observation of a constant isoelective temperature is one requirement of a common correlation of relative reactivities with substituent effects; that all of the reactions proceed by the same mechanism is another.

Ab initio calculations have shown that exothermicity in singlet carbene cycloadditions to ethylene increases with increasing electron donation by the carbene substitutent.[78] The more exothermic the reaction (e.g., $\Delta E_{rxn\,2}^{CCl} > \Delta E_{rxn}^{CF_2}$), the lower the activation energy, and the "earlier" the transition state. The lower the activation energy, the lower the selectivity. Calculated values of E_{rxn} are very close to thermochemical estimates.[78]

B. Carbene Philicity

The electrophilic character of experimentally generated carbenes has been amply demonstrated.[5-9] The observed correlations defined by Equations 10 to 12 together with that between m_{CXY}^{calcd} values and carbene LUMO energies[78] fortify this interpretation. Electron-withdrawing

groups lower carbene LUMO energies and decrease carbene selectivity toward addition to the normal set of alkenes employed for m_{CXY} determinations, whereas electron-donating substituents raise carbene LUMO energies and increase carbene selectivity.[14] One might have expected to find carbenes approaching the inner (diffusion control) and outer (stable) limits of reactivity with suitable changes in carbene substituents.

As early as 1974, dimethoxycarbene (m_{CXY}^{calcd} = 2.22) was observed to have unusual reaction characteristics.[79] Addition to simple alkenes like cyclohexene did not occur, but cyclopropane formation was observed in its reactions with styrene, diethyl maleate or fumarate, and ethyl cinnamate.[79,80] Preferential addition to "electrophilic" olefins defined $(MeO)_2C$ as a nucleophilic carbene. With even greater electron donation provided by the dimethylamino group, $MeOCNMe_2$ (m_{CXY}^{calcd} = 2.91) is even less reactive toward addition to the carbon-carbon double bond.[81]

Between the nucleophilic $(MeO)_2C$ and the characteristically electrophilic CF_2 lie alkoxyhalocarbenes that have both nucleophilic and electrophilic reactivities and are, consequently,

Table 9
RELATIVE REACTIVITIES OF REPRESENTATIVE ELECTROPHILIC AND AMBIPHILIC CARBENES TOWARD OLEFINS

	Relative reactivity			
Olefin	MeOCCl[a]	PhOCCl[b]	CCl_2[c]	MeCCl[d]
$Me_2C=CMe_2$	12.6	3.0	78.4	7.44
$Me_2C=CH_2$	5.43	7.3	4.89	1.92
t-MeCH=CHMe	1.00	1.00[e]	1.00	1.00
$H_2C=CH(CH_2)_3CH_3$		0.36		
$H_2C=CHCOOMe$	29.7	3.7	0.060	0.078
$H_2C=CHCN$	54.6	5.5	0.047	0.074
m_{CXY}^{calcd}	1.59	1.49	0.97	0.58

[a]   Reference 83, 25°C.
[b]   Reference 84, 25°C.
[c]   Reference 87, 80°C.
[d]   References 53 and 87, 25°C.
[e]   The standard olefin is *trans*-2-pentene instead of *trans*-2-butene.

referred to as *ambiphiles*. The ambiphilic reactivities of two carbenes, MeOCCl and PhOCCl, are well defined.[82-86] Relative reactivities for these carbenes display a "parabolic" dependence on alkene ionization potential, reacting rapidly with both "electron-rich" tetramethylethylene and "electron-poor" acrylonitrile (Table 9). In contrast, electrophilic carbenes such as CCl_2 and MeCCl display steadily decreasing reactivities as the π-electron-donating abilities of the alkene substrates decrease.[87] Furthermore, methoxychlorocarbene, in contrast to CCl_2 and CF_2, exhibits a nonlinear Hammett correlation in its addition to a series of substituted styrenes: p-MeO, 1.50; p-CH_3, 1.07; H, 1.00; m-Cl, 1.04; m-NO_2, 1.27.[85]

Equation 10 has been employed to provide calculated measures of carbene reactivities for both ambiphilic and nucleophilic carbenes where experimental values are often unavailable. However, data such as that presented in Table 9 suggest that Equation 10 may be relevant only to electrophilic carbenes. The m_{CXY}^{calcd} value for MeOCCl (1.59) leads to the prediction that the spread in relative reactivities for addition to olefins ranging from tetramethylethylene to *trans*-2-butene should exceed that for CF_2 ($m_{CXY}^{calcd} = 1.47$) and certainly be much greater than that for CCl_2 ($m_{CXY}^{calcd} = 0.97$). Clearly, this is not the case and, although differences can be attributed, at least partly, to steric effects, the net result is that the selectivity of MeOCCl as an electrophile is much less than predicted from its m_{CXY}^{calcd} value.

A rationalization for decreased selectivity in reactions of ambiphilic carbenes toward olefins has been provided recently. Moss employed substrate orbital energies and calculated values for the HOMO (σ) and LUMO (p) orbital energies of MeOCCl, CCl_2, and CF_2 to obtain differential orbital energies, $E = (E_{CXY}^{LU} - E_{XSty}^{HO})$ and $N = (E_{XSty}^{LU} - E_{CXY}^{HO})$ for the frontier molecular orbital interactions of each carbene with a series of substituted styrenes (XSty).[85] For reactions of CCl_2 and CF_2, $E < N$ for the complete set of substituted styrenes. However, for reactions of CH_3OCCl, which displayed a minimum in relative reactivity when $X = H$, $E < N$ when X was an electron-donating substituent, and $E > N$ when X was an electron-withdrawing substituent. Moss has argued that since both sets of orbital interactions (E and N) contribute to transition state stabilization for carbene addition to olefins, the spread in relative reactivities decreases when $E \simeq N$.[85]

Ab initio calculations have located the transition state for the cycloaddition of singlet carbenes to ethylene.[78] Four carbenes, CCl_2, CF_2, HOCF, and $C(OH)_2$, corresponding to

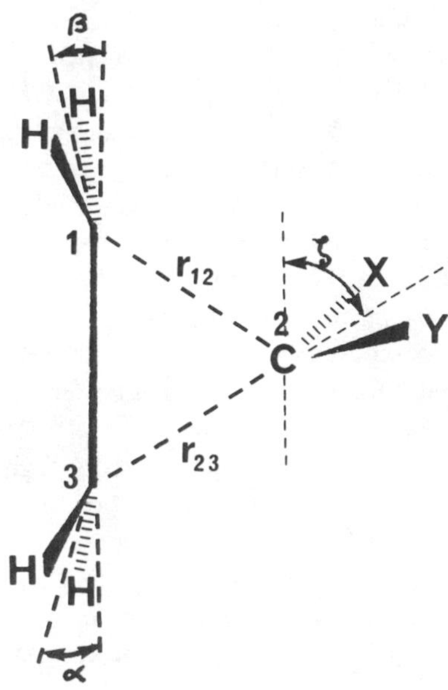

FIGURE 2.   Transition state structures of carbene cycloadditions with ethylene.

Table 10
TRANSITION STATE STRUCTURAL PARAMETERS FOR CARBENE CYCLOADDITIONS WITH ETHYLENE (FIGURE 2)[78]

Carbene	r_{12}, Å	r_{23}, Å	r_{12}/r_{23}	ζ, deg.	α, deg.	β, deg.
CCl_2	2.311	1.958	1.1803	36	13.9	1.9
CF_2	2.220	1.801	1.2326	43	22.2	1.3
HOCF	2.350	1.776	1.3232	48	25.0	2.6
$C(OH)_2$	2.227	1.707	1.3046	58	29.7	4.2

electrophilic, ambiphilic, and nucleophilic carbene systems, were examined, and the following conclusions were reached: (1) the ratio of newly forming bond lengths (r_{12}/r_{23} in Figure 2) increases along the series CCl_2, CF_2, HOCF, indicating greater asymmetry in bond formation until finally with $C(OH)_2$ this ratio begins to decrease as the transition state becomes more cyclopropane-like (Table 10). (2) The angle of tilt (ζ) of the CCl_2 and CF_2 planes with respect to the original ethylene plane indicates slightly greater π than σ bonding, whereas ζ for the $C(OH)_2$ transition state suggests a predominantly σ, or nucleophilic, interaction with ethylene in the transition state. The angle ζ (Figure 2) is expected to be 0° for a purely electrophilic interaction and 90° for a purely nucleophilic interaction. (3) The increased angle of distortion away from planarity of the ethylene CH_2 groups, which is pronounced in α, also indicates of increased nucleophilic character in the series CCl_2 < CF_2 < HOCF < $C(OH)_2$ (α = 30° in cyclopropane).

Frontier molecular orbital treatment of carbene cycloadditions, which assesses the relative importance of the carbene LUMO-alkene HOMO interaction (favors a π or electrophilic, transition state) and the carbene HOMO-alkene LUMO interaction (favors an unsymmetrical σ, or nucleophilic, transition state), has provided a purely theoretical index for carbene philicity, PI_{CXY}.[78] Based in second-order perturbation theory, PI_{CXY} is defined as the ratio of stabilization energies arising from nucleophilic and electrophilic interactions (PI_{CXY} =

E_n/E_e). PI_{CXY} values (Table 5) that are less than 1.0 ($ClCCH_3$, FCPh, $ClCSCH_3$, CCl_2, FCCl, and CF_2) imply that the carbene is electrophilic. Carbenes whose PI_{CXY} values fall between 1.0 and 1.2 are ambiphilic ($ClCOCH_3$, $FCOCH_3$, and FCOH). Nucleophilic carbenes have PI_{CXY} values greater than 1.2 ($C(OMe)_2$, $C(OH)_2$, and $MeOCNMe_2$).

Theoretical predictions correspond to those determined experimentally,[14] and the conclusion that the stabilization energy of the carbene is due to resonance electron donation from the substituents to the vacant p orbital of the carbene has been drawn from this correspondence.[78] However, minimization of the importance of the inductive term is not justified for carbene cycloadditions that are performed much above 25°C.[75] Furthermore, the consensus is undecided about the relative importance of steric effects in these carbene addition reactions.

The complexity of carbene philicity has been suggested from studies of thermal decomposition of 3-chloro-3-phenyldiazirine in the presence of olefins ranging from vinyl ethers to α,β-unsaturated carbonyl compounds.[88] Unexpectedly high yields of cyclopropane products were obtained, especially by comparison with those from decomposition of 3-chloro-3-methoxydiazirine in the presence of α,β-unsaturated carbonyl compounds and nitriles.[82,89,90] Of special significance were results from the decomposition of 3-chloro-3-phenyldiazirine in the presence of diethyl fumarate and diethyl maleate (Equations 14 and 15). With diethyl fumarate only one cyclopropane product was produced, and its structure

$$\text{(14)}$$

$$\text{(15)}$$

retained the *trans* geometry of the carboethoxy groups. However, reaction with diethyl maleate under the same conditions produced three cyclopropane products: two that retained the *cis* geometry of the reactant olefin and one in which the carboethoxy groups were *trans*. Although product yields were low with diethyl maleate (16% when the reactant ratio was 3.0), the formation of the product with *trans* carboethoxy groups could not be attributed to the amount of diethyl fumarate present in the maleate reactant or to isomerization of diethyl maleate. This rearrangement demands the intervention of a reaction intermediate with sufficient lifetime to allow bond rotation to occur. Consideration was given to thermal isomerization of the diazirine to the diazo compound followed by dipolar addition and dinitrogen extrusion. According to this mechanism, the isomerization step must be rate limiting since the rate for dinitrogen loss was independent of the olefin employed.[88] However, cyclopropane production from reactions with α,β-unsaturated esters and nitriles occurred without competition with formal vinyl C–H insertion which is commonly observed from 1-pyrazoline decomposition.[39,91] Furthermore, the addition of bases normally effective in catalyzing the tautomerization of 1-pyrazoline to stable 2-pyrazoline products[38] only caused a decrease in the yield of cyclopropane, but not of dinitrogen, presumably due to base association with the intermediate carbene.

Doyle and co-workers have advanced an alternative pathway involving association of the normally electrophilic carbene with a Lewis base resulting in the production of a nucleophilic ylide species (Scheme 3).[88]

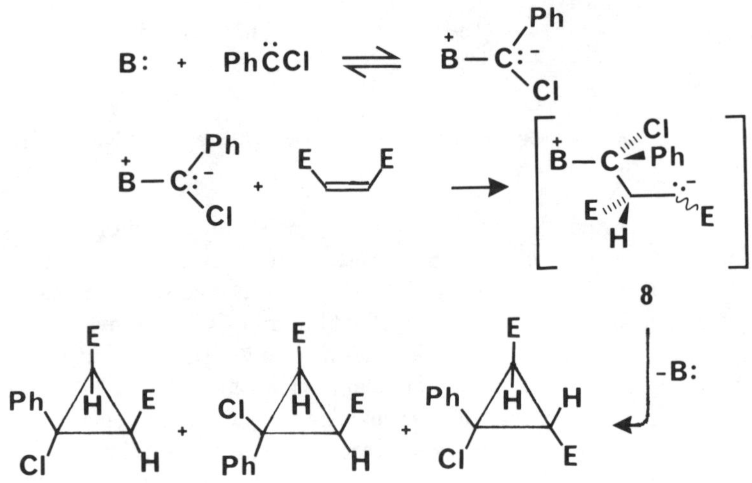

SCHEME 3.

Nucleophilic addition of the ylide to the carbon-carbon double bond of a conjugated ester or nitrile can be expected to produce a reaction intermediate (**8**) capable of isomerization as a result of bond rotation.[92] Ylide formation resulting from carbene interaction with the unshared electron pairs of heteroatoms is well-established with simple carbenes such as CH_2[93,94] and with α-carboalkoxy or α-ketocarbenes and dihalocarbenes.[95,96] However, more recent investigations have established that B: in Scheme 3 is not the olefinic reactant, because relative reactivity studies using varying ratios of "electrophilic" and "nucleophilic" olefins show a similar order of dependence on the concentrations of both olefins.[97] Furthermore, relative reactivities obtained from these olefin pairs are only moderately sensitive to iodobenzene, even at a molar ratio of iodobenzene to 3-chloro-3-phenyldiazirine of 100:1 or more, and actual values for relative reactivities (diethyl fumarate vs. cyclohexene and n-butyl vinyl ether vs. ethyl acrylate) change by less than 30%. This latter result was surprising since iodobenzene has recently been shown to be an effective trap for certain carbenes.[98]

C. Carbene-Alkene Complexes

Considerable controversy surrounds the possible intervention of carbene-alkene complexes in addition reactions. Skell and Garner[43] first proposed the mechanism in which the vacant p orbital of the singlet carbene approaches the olefin along the π axis closer to the less substituted carbon atom to form a loose charge-transfer type complex which then collapses to the product cyclopropane. This mechanism was further refined by Hoffmann,[99] and it was generally accepted that intermediate π-complexes or "loose" charge-transfer complexes are involved in carbene addition reactions. However, Houk described *ab initio* calculations which suggest that the most reactive halocarbenes do not form stable π-complexes with alkenes.[100,101]

Table 11
ABSOLUTE RATE CONSTANTS FOR
CYCLOADDITIONS OF
PHENYLHALOCARBENES TO ALKENES[a]

Alkene	$k_{carbene}, M^{-1} sec^{-1}$		
	PhCF	PhCCl	PhCBr
$Me_2C=CMe_2$	1.6×10^8	2.8×10^8	3.8×10^8
$Me_2C=CHMe$	5.3×10^7	1.3×10^8	1.8×10^8
t-MeCH=CHEt	2.4×10^6	5.5×10^6	1.2×10^7
n-BuCH=CH$_2$	9.3×10^5	2.2×10^6	4.0×10^6
m_{CXY}	0.89	0.83	0.70

[a]   Reactions performed in alkene/isooctane solution at 23°C (Reference 102).

Calculations were carried out at the 3-21G level on the cycloadditions of CCl_2 and CF_2 to ethylene, and intermediate complexes consisting of an essentially undistorted ethylene arranged in a plane nearly parallel to that of the slightly distorted singlet carbene were found. The energy of the transition structure for cycloaddition of CCl_2 to ethylene was lower than that of the reactants, so that a negative activation energy, apparently in good accord with the experimental results on PhCCl[102,103] and inferences about CCl_2,[46,68,75] was predicted. However, the complex disappears with the inclusion of correlation energy, and no barrier at all to cycloaddition was predicted from these calculations. Similar results emerged for cycloaddition to isobutylene.[100]

In contrast to calculations for CCl_2, the more selective CF_2 does form a complex with ethylene according to *ab initio* calculations at the levels explored by Houk et al.[100] The activation energy for this addition is relatively large (26.9 kcal/mol from 3-21G calculations), and a weakly bound complex ($\Delta E = -2.0$ kcal/mol) can be predicted. However, Houk argues that ether or halocarbon solvents often used for carbene addition reactions should be superior Lewis base "coordinators" than alkenes, and he cites calculated energies of -23.6 and -9.7 kcal/mol for complexes of CCl_2 and CF_2, respectively, with water.

Moss and co-workers[102-105] have obtained absolute rate constants for cycloadditions of phenylhalocarbenes with representative alkenes (Table 11). Although these rate constants are system dependent,[103] varying by factors as high as 2 to 3 using different laser flash systems, comparative results have provided useful precise information. Temperature dependencies have been determined for reactions of PhCCl with these olefins, and negative activation energies were observed for two of the four reacting systems:[103] $Me_2C=CMe_2$ ($E_a^{obsd} = -1.7 \pm 0.5$ kcal/mol), $Me_2C=CHMe$ ($E_a^{obsd} = -0.77 \pm 0.5$ kcal/mol), t-MeCH=CH$_2$ ($E_a^{obsd} = 1 \pm 0.5$ kcal/mol). Using the "standard" interpretation for negative activation energies — that the reaction is a multistep process that involves at least one intermediate which possesses at least two "channels" for reaction — the authors interpreted their results in terms of a mechanism involving a reversibly formed dissociable intermediate (Equations 16 and 17).[68]

$$PhCCl + A \underset{k_{-1}}{\overset{k_1}{\rightleftharpoons}} PhCCl/A \tag{16}$$

$$PhCCl/A \overset{k_2}{\rightarrow} Cyclopropane \tag{17}$$

Table 12
DIFFERENTIAL ACTIVATION PARAMETERS FOR
CYCLOADDITION OF PhCCl WITH ALKENES[103]

Alkene	$\Delta\Delta G^{\dagger}$ (kcal/mol)	$\Delta\Delta H^{\dagger}$ (kcal/mol)	$\Delta\Delta S^{\dagger}$ (eu)
$Me_2C{=}CMe_2$	2.2	-4.4	-23
$Me_2C{=}CHMe$	2.7	-3.7	-22
$t\text{-MeCH}{=}CHEt$	4.3	-1.6	-20
$H_2C{=}CH(CH_2)_3CH_3$	4.5	-1.5	-21

Consequently, $k_{obsd} = k_1 k_2/(k_{-1} + k_2)$ so that, since k_1 corresponds to normal diffusion (at temperature T in isooctane), the ratio of rate constants k_1/k_2 is available from

$$k_{-1}/k_2 = (k_1 - k_{obsd}/k_{obsd}) \tag{18}$$

Values of $\Delta\Delta G^{\dagger}$, $\Delta\Delta H^{\dagger}$, and $\Delta\Delta S^{\dagger}$ for the competition between dissociation and addition, computed from the plot of log (k_{-1}/k_2) vs. $1/T$ (Table 12),[103] suggested that the differences in rates for the reactions of PhCCl with alkenes at any given temperature are due to changes in the relative enthalpies for cyclopropanation vs. dissociation.

Houk and Rondan[101] have argued that negative activation energies and entropy control of reactivity can arise in fast reactions having no inherent potential energy barrier. In addressing Giese's arguments for entropy control in the π-complex model for cycloaddition,[68,74,75,77,106,107] they find that the more substituted alkene has a more favorable entropy because the transition state occurs earlier. Geise had concluded from a detailed analysis of the dependencies of differential enthalpic ($\Delta\Delta H^{\dagger}$) and entropic ($\Delta\Delta S^{\dagger}$) terms on substrate structure and carbenic substituents that despite inherent limitations of the isokinetic relationship,[5,108] entropic effects dominate the selectivities of CBr_2, $CBrCl$, and CCl_2 cycloadditions.[74,107] Furthermore, he has stated that the effects of methoxy and phenyl substituents on alkene reactivity operate mainly on k_2 (Equation 17), rather than on the complex formation equilibrium (Equation 16).[75] Houk and Rondan concluded that the temperature effects upon the kinetic behavior of reactive species are dictated by how fast the "walls of the entrance channel" on the potential energy surface narrow. For reactions with no significant barriers, the transition state occurs when the decrease in ΔH overcomes the unfavorable entropy effects, and $\Delta H^{\dagger}$ is negative even when a complex is not formed. Diffusion-controlled reactions occur when $\Delta G^{\dagger}$ (diffusion) exceeds the $\Delta G^{\dagger}$ for reaction, and reactions that are diffusion-controlled at some temperature should develop negative activation energies at higher temperatures, as has been reported for reactions of PhCCl with 2,3-dimethyl-2-butene and with 2-methyl-2-butene.[103] Houk and Rondan[101] predicted that reactions of CCl_2, $CBrCl$, and CBr_2 will be unselective and diffusion controlled at low temperature. In contrast, the Giese π-complex model predicts that carbene selectivities will reverse at low temperatures, with alkyl substitution eventually diminishing the rate of reaction.[107,108] The answer to these different predictions should be forthcoming.

Several carbene cycloadditions have resulted in abnormal products that suggest the intervention of carbene-alkene complexes: (1) the addition of CHX (X = Cl,I) to 1,2-dimethylcyclobutene (Equation 19) which produces 2-(1-methylcyclopropyl)-1-halopropenes **9** and not the expected 5-halo-1,4-dimethylbicyclo[2.1.0]pentanes,[109]

CH$_3$ / CH$_3$ + :CHX (from I$_2$CHX, hν) $\longrightarrow$ H$_3$C, H, X, CH$_3$ (**9Z**) + H$_3$C, X, H, CH$_3$ (**9E**)

$$\tag{19}$$

Table 13
PRODUCT DISTRIBUTION FROM REACTIONS OF
$(CF_3)_2C$ WITH 2-BUTENE[110]

Alkene	Carbene precursor[a]	10c	10t	11	12[b]	13[b]
trans-2-butene	$(CF_3)_2C=N_2$, 150°C	0	100	—	—	—
	c-$(CF_3)_2CN_2$, 165°C	0	57	0	39	4
cis-2-butene	$(CF_3)_2C=N_2$, 150°C	39	8	49	—	—
	c-$(CF_3)_2CN_2$, 165°C	55	9	2	27	8

[a] c-N_2 denotes diazirine.
[b]   Apparent insertion occurs stereospecifically with preservation of olefin geometry.

(2) rearrangement accompanying the addition of $(CF_3)_2C$ to *cis*- and *trans*-2-butene resulting in **11** (Table 13),[110] and (3) isomerization of *trans*-cyclooctene during

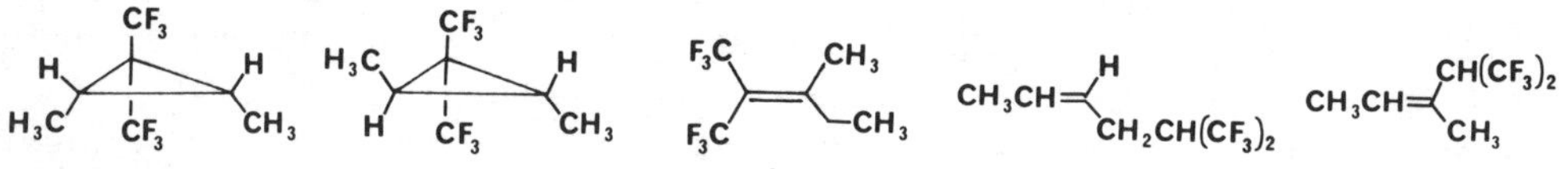

dihalocarbene addition (Equation 20, X = Cl, Br).[111,112]

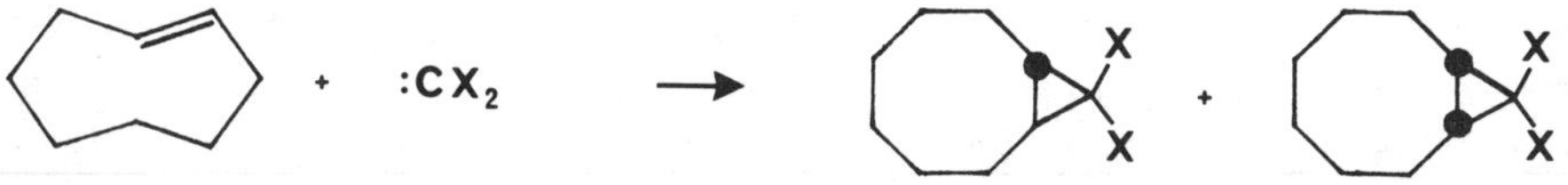

$$X = Cl \quad 100 \quad 0$$
$$Br \quad 77 \quad 33 \tag{20}$$

Although CHI, generated photolytically from iodoform, underwent addition to *cis*-2-butene without rearrangement, **9** was formed exclusively ($E/Z = 1.0$) from 1,2-dimethylcyclobutene and CHI or CHCl, and the dipolar species **16** was suggested as an intermediate in the carbene addition reaction. Further studies on this interesting system have not been reported, but they are certainly warranted.

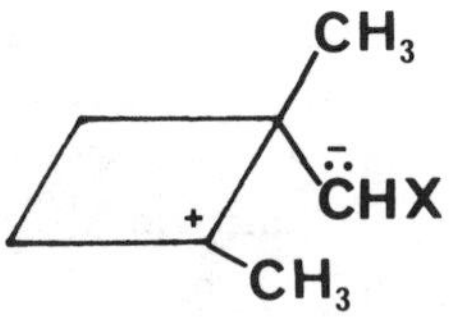

The production of **11—13** in reactions of $(CF_3)_2C$ with the isomeric 2-butenes (Table 12) was suggested to occur by the intervention of a diradical intermediate.[110] Accordingly, insertion products **11** and **13** could be derived from intermediate **17**

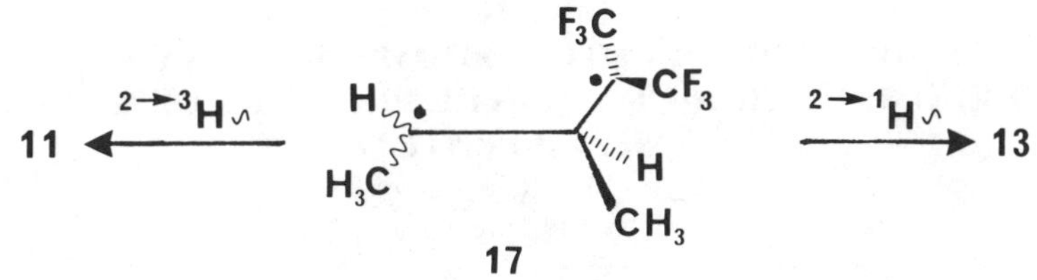

$$\textbf{11} \quad \xleftarrow{\;2 \to 3\,H\curvearrowleft\;} \quad \textbf{17} \quad \xrightarrow{\;2 \to 1\,H\curvearrowleft\;} \quad \textbf{13} \tag{21}$$

by hydrogen transfer rather than by direct C–H insertion. A similar intermediate explains the apparent insertion products formed by reactions of cyclohexane with either $(CF_3)_2C$[110] or PhCCl (Equation 22).[88,113] Moss reported a product resembling **11** from the reaction of CF_3CCl, generated from $PhHg(CF_3)CClBr$ at 130°C, and suggests that this olefin arises from thermal rearrangement of the cyclopropanation product.[58] Appropriate control and/or labeling experiments have yet to be performed on such systems to differentiate the most probable pathway.

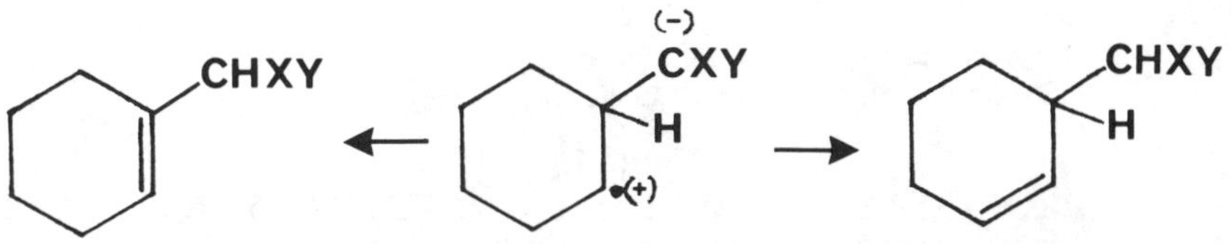

$$\tag{22}$$

Generation of CBr_2 in the presence of *trans*-cyclooctene resulted in isomerization of *trans*- to *cis*-cyclooctene.[111,112] In contrast, CCl_2 underwent stereospecific cycloaddition with *trans*-cyclooctene. The results were interpreted[112] in terms of an intermediate complex between the cycloalkene and carbene which caused isomerization as a result of the energy released upon carbene-alkene complex dissociation. A diradical intermediate was ruled out on the basis that 1,1-dicyclopropylethylene underwent addition with CBr_2 without rearrangement.[114] Likewise, triplet CBr_2 is not involved in reactions of this type.[115] However, a dipolar species similar to **16** may be in accord with the available data.

Each of the systems described in Equations 19 and 20 has a built-in tripping lever to signal a nonconcerted disruption of the π-bond in the alkene undergoing cycloaddition. This approach is intriguing, but only a limited amount of data are currently available. Carbenes generated from diazirines, either thermally or photochemically, have not been employed with these alkenes, and further research in this area is justified.

Liu et al. have evolved a subtle alternative approach to resolve the questions of carbene-alkene complex involvement in carbene cycloaddition reactions.[116-119] Upon photolysis or thermolysis, 3-benzyl-3-chlorodiazirine undergoes dinitrogen extrusion and subsequent intramolecular 1,2-hydrogen migration in direct competition with intermolecular cycloaddition to alkenes (Scheme 4).

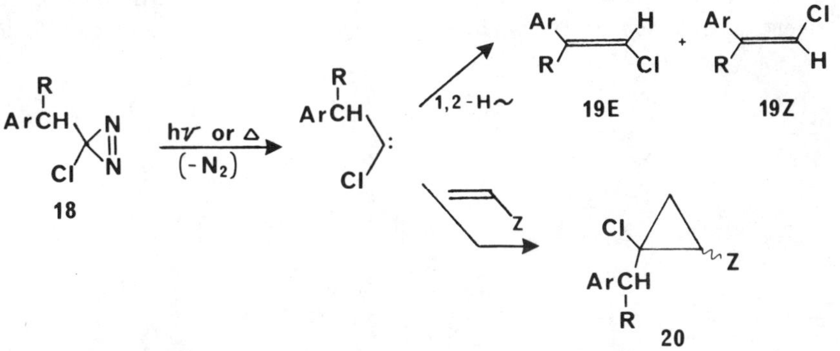

SCHEME 4.

Table 14
PRODUCT DISTRIBUTION FROM PHOTOLYTIC DECOMPOSITION OF 18 IN THE PRESENCE OF (Z)-4-METHYL-2-PENTENE[a]

18		19E/19Z for [alkene]/[23] =					
Ar	R	0	10	100	500	∞[b]	19/20[c]
p-CH$_3$OC$_6$H$_4$	H	15.2	14.2	10.2	8.2	6.8	4.93
p-CH$_3$C$_6$H$_4$	H	11.9	11.5	7.0	4.6	3.9	6.62
C$_6$H$_5$	H	5.7	5.0	3.1	2.1	1.9	6.88
3,4-Cl$_2$C$_6$H$_3$	H	5.4	4.9	3.8	4.3	4.2	3.37
p-ClC$_6$H$_4$	H	5.3	4.5	3.2	2.5	2.3	5.49
C$_6$H$_5$	CH$_3$	3.5	—	3.4	3.3	2.9	0.47

[a] Reactions performed at 10°C with irradiation through a Corning CS-052 filter.

[b] Neat solution of alkene.

[c] Ratios from reactions performed in neat (Z)-4-methyl-2-pentene.

Table 15
PRODUCT DISTRIBUTION FROM PHOTOLYTIC DECOMPOSITION OF 18 IN THE PRESENCE OF REPRESENTATIVE ALKENES[a]

Alkene[b]	18: Ar=Ph, R=Me		18: Ar=Ph, R=H	
	19E/19Z	20/19	19E/19Z	20/19
1-Hexane	3.24	0.21	2.40	4.85
(Z)-4-Methyl-2-pentene	2.91	0.47	1.88	6.88
2-Methyl-2-butene	2.55	0.50	1.50	6.41
Diethyl fumarate[c]	—	—	(3.70)	(0.51)

[a] Reactions performed at 10°C with irradiation through a Corning CS-052 filter.

[b] Neat alkene.

[c] Reaction performed at 8.5°C with [alkene]/[18] = 60.

The 1,2-hydrogen shift produces both (E)- and (Z)-β-chlorostyrene in a ratio that is dependent on the structure of the reactant diazirine and on the concentration of the alkene (Table 14),[117,118] as well as on the structure of the alkene (Table 15)[117,119] and on the reaction temperature.[119] As the data in Table 14 suggest, no apparent structure-reactivity relationship for either **19E/19Z** or **20/19**, and a plot of **20/19** as a function of alkene concentration shows pronounced curvature.[117] However, if these results are interpreted in terms of an intermediate carbene-alkene adduct (Scheme 5), where the free carbene and/or its adduct with an alkene

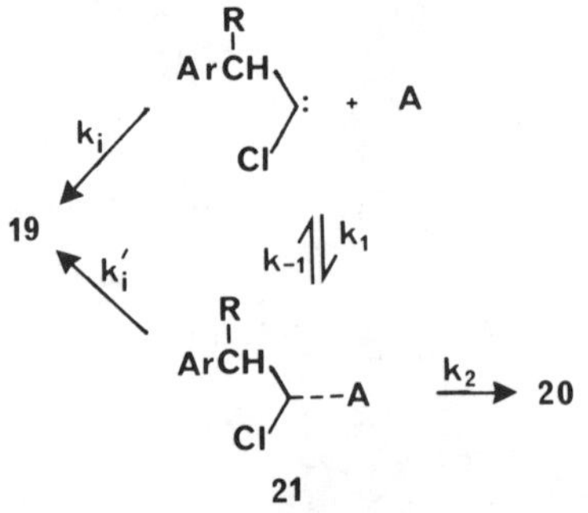

SCHEME 5.

(21) undergoes intramolecular 1,2-hydrogen migration, but only the adduct forms cyclopropane, Scheme 5 predicts an inverse dependence of **19/20** on the alkene concentration:[117,119]

$$\frac{19}{20} = \frac{k'_i}{k_2} + \frac{k_i}{k_t} \cdot \frac{1}{[\text{alkene}]}$$

where $k_t = k_1 k_2 / (k_{-1} + k_2 + k_i)$. This inverse dependence is observed, and the detection of substrate dependence in the **19E/19Z** ratio seems to exclude the alternative excited carbene explanation proposed by Warner.[120]

1,2-Hydrogen migration occurs through two limiting conformers[118] **22** and **23** (Scheme 6).

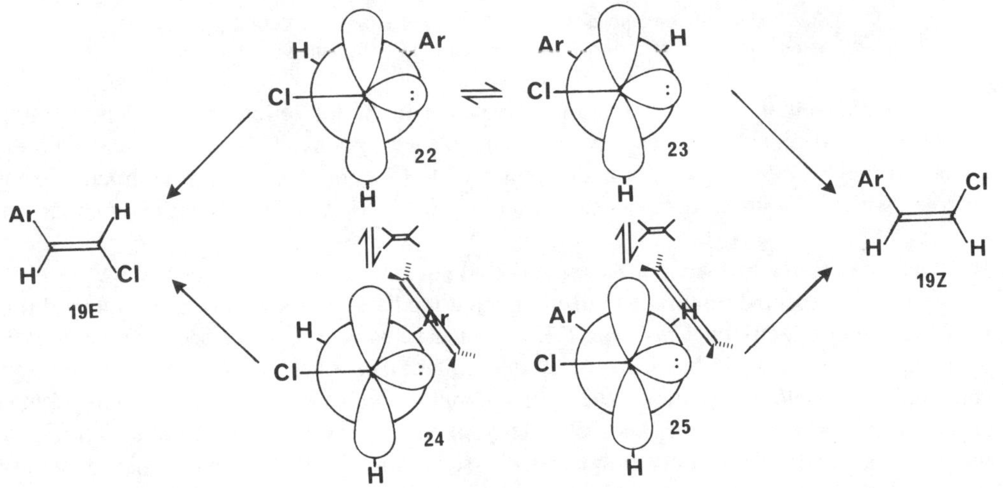

SCHEME 6.

β-Chlorostyrene **19E** arises from **22**, and **19Z** is formed from **23**. Interaction of the carbene with an olefin should not be a simple π-coordination with the carbene LUMO because such an explanation fails to account for the effect of olefin structure and concentration on the stereochemical outcome of the 1,2-hydrogen migration (Tables 14 and 15). Instead, the olefin may be properly oriented in the bisected form depicted in Figure 2. As a result of olefin repulsion by the aryl group, **25** is expected to be more stable than **24**, and this explanation accounts for the increase in **19Z** with increasing olefin concentration. The lack of any substantial influence of olefin structure or concentration on the **19E/19Z** ratio with chloro(1-phenylethyl)carbene (from **18**: Ar=Ph, R=Me) is also in accord with Scheme 6.

The depictions of **24** and **25** are not intended to imply a specific π-complex. Rather, these structures are meant to suggest a spectrum of intermediates whose precise formulation is dependent on the carbene and olefin. A σ-complex would explain structural rearrangements (e.g., Equations 19 and 20) that can accompany "normal" modes of reaction. Such complexes may not be involved in all carbene reactions. However, they do offer a satisfying explanation for a diverse set of experimental results that cannot be explained by concerted cycloaddition.

Recent examination of the reactions of chlorocarbenes with diethyl fumarate[119] has suggested that Scheme 5 is in accord with addition of an electrophilic carbene to an electron-deficient olefin. Absolute rate constants for $p\text{-NO}_2\text{C}_6\text{H}_4\ddot{\text{C}}\text{Cl}$, $\text{C}_6\text{H}_5\ddot{\text{C}}\text{Cl}$, and $p\text{-CH}_3\text{OC}_6\text{H}_4\ddot{\text{C}}\text{Cl}$ with diethyl fumarate are 1.0×10^8, 3.8×10^6, and 6.3×10^5 M/sec, respectively, and diethyl fumarate ($k_t/k_i = 1.40$ at 25°C) is less reactive than tetramethylethylene ($k_t/k_i = 10.4$ at 24°C), which are in agreement with an electrophilic carbene addition. Furthermore,

Table 16
STEREOSELECTIVITIES FOR CARBENES GENERATED
PHOTOLYTICALLY FROM DIAZIRINES AT 25°C

Carbene	$m_{CXY}^{25°C}$	(CH$_3$)$_2$C=CHCH$_3$	*cis*-CH$_3$CH=CHCH$_3$	Ref.
			syn-X/*syn*-R	
CF$_3$CCl	0.19	1.48	1.65	58
CpCCl	0.41	—	4.6	122
MeCCl	0.50	1.45	2.84	53
PhCBr	0.70	1.31 (1.28)[a]	1.55 (1.35)[a]	52
PhCCl	0.83	1.29 (1.18)[b]	1.97 (2.10)[b]	51

[a]   Stereoselectivity for carbene generated from PhCHBr$_2$ with KO-*t*-Bu.
[b]   Stereoselectivity for carbene generated from PhCHCl$_2$ with KO-*t*-Bu.

E_t for diethyl fumarate is 1.8 kcal/mol greater than E_t for tetramethylethylene.[121] It is conceivable that dimethyl maleate, whose reaction with phenylchlorocarbene has been explained through intervention of a nucleophilic ylide (Scheme 3),[88] forms a carbene-alkene complex which, like *trans*-cyclooctene (Equation 20), is triggered to undergo isomerization.

D. Stereoselectivity in Carbene Cycloaddition

As could be expected from such high energy intermediates, carbenes undergo cycloaddition with low stereoselectivity. This aspect of carbenic selectivity has not been investigated in great detail, and comparable data are quite limited. However, stereoselectivity data are commonly available for additions to 2-methyl-2-butene and *cis*-2-butene for the temperature at which the carbene is generated. Table 16 summarizes results from carbene additions to these olefins under identical conditions. Stereoselectivities for PhCCl and PhCBr, generated from benzal chloride and benzal bromide, respectively, differ, but not substantially, from those obtained by photolysis of diazirines.

Only small changes in isomer ratios are ordinarily observed with changes in carbene or olefin substituents. Among those olefins for which comparisons of stereoselectivities are possible, 2-methyl-2-butene appears to undergo addition with selectivities that are proportional to m_{CXY}, whereas *cis*-2-butene results suggest no such relationship (Table 16). By themselves, these differences in stereoselectivity are too small to be mechanistically meaningful. However, in a uniform series they do offer insight into the steric and electronic factors that govern these cycloaddition reactions.[58,122]

The relatively high stereoselectivity observed for the reaction of CpCCl with *cis*-2-butene has been offered as further evidence for preferential addition by the "twisted" conformer **6t**.[122] Although the bisected conformer **6b** is more stable than **6t** by 9.5 kcal/mol,[54] **6b** encounters substantial steric hindrance from cyclopropyl interactions with the olefin methyl groups. However, steric hindrance is largely absent if addition occurs with **6t** so that the olefin methyl groups are syn to the carbene chloride substituent.

The stereochemistry of carbene additions to vinyl ethers has recently been examined.[88,123,124] These reactions have synthetic value, because the cyclopropane intermediate is susceptible to thermal rearrangement (e.g., Equations 23[124] and 24[125]). Only small stereochemical preferences are observed from addition of either methylchlorocarbene or methylbromocarbene, generated from the corresponding 1,1-dihaloethanes with methyllithium/lithium halide, to a series of acyclic vinyl ethers.[123] However, these results have been interpreted in terms of carbenoid intermediates.

Table 17
REGIOSELECTIVITIES IN
CYCLOPROPANATION OF 1,3-DIENES

Diene	3,4-addition/1,2-addition	
	$:CCl_2$	$Rh_2(OAc)_4$, EDA[a]
$H_2C=C(CH_3)CH=CH_2$	0.020[b]	0.63
$H_2C=C(Ph)CH=CH_2$	<0.01[a,b]	0.44
$H_2C=C(Cl)CH=CH_2$	0.053[d]	0.26
$H_2C=CHCH=CHCH_3$	1.4[c]	7.5
$H_2C=CHCH=CHPh$	16[a,e]	50.0
$H_2C=CHCH=CHCl$	>20[f]	>100.0

[a] EDA = ethyl diazoacetate. Data from Reference 126.
[b] Data from Reference 127.
[c] Data from Reference 128.
[d] Data from Reference 129.
[e] Data from Reference 130.
[f] Data from Reference 131.

$$(23)$$

$$(24)$$

E. Regioselectivity in Carbene Cycloaddition

Reactions of carbenes with dienes and polyenes have received relatively little attention. With conjugated dienes, 1,2-addition is the general rule, although these products do undergo thermal vinylcyclopropane-cyclopentene rearrangement. Dichlorocarbene reactions with a series of substitutes 1,3-butadienes have been reported (Table 17) and regioselectivities for dibromocarbene addition to the same set of dienes, where known,[127] are slightly greater. However, product instability with CBr_2 addition products has limited thorough analyses.

The data reported in Table 17 for CCl_2 addition to 1,3-dienes contrast with results obtained for catalytic cyclopropanation with an electrophilic carbene.[126] 1-Substituted-1,3-butadienes undergo preferential cyclopropanation at the 3,4-double bond, and regioselectivities from CCl_2 addition parallel those from rhodium(II) acetate catalyzed cyclopropanation with ethyl diazoacetate. Reactions of 2-substituted-1,3-butadienes with ethyl diazoacetate catalyzed by $Rh_2(OAc)_4$ show a gradual predictable[132] increase in selectivity with substituents Me < Ph < Cl. However, chloroprene is less selective than isoprene in reactions with CCl_2. Since the selectivity for addition to the 1,2-double bond of chloroprene would have been expected to be 5 to 10 times greater than observed, this result suggests significant electronic repulsion between chloro substituents on the carbene and diene in the transition state for addition.

Carbene addition with dienes occurs with a much greater preference for reaction at the

1,2-double bond of 2-substituted-1,3-butadienes than at the 3,4-double bond of 1-substituted-1,3-butadienes. A similar preference is reported for epoxidation reactions,[127] but the reverse is observed for reactions with metal carbenes[126] and in Diels-Alder reactions.[133] The dominant influence of substituents directly attached to the reaction center may be characteristic of an early transition state, and additional data is warranted.

Rondan and Houk[134] have recently reported the transition structures for the reaction of difluorocarbene with propene. They conclude that although the activation energy for addition of CF_2 to propene is 1.3 kcal/mol lower than the addition of CF_2 to ethylene, the average transition structure for the propene reaction is essentially the same as that for the ethylene reaction. Furthermore, the activation energy for CF_2 cycloaddition to propene via the transition state where the carbene fluoro-substituents are anti to the methyl group of the olefin does not differ from that of the syn transition state. This leads to the conclusion that unsymmetrical alkyl substitution on the olefin has no appreciable influence on which end of the double bond is approached preferentially by the attacking difluorocarbene. The high regioselectivities observed for cycloaddition of CCl_2 to 2-substituted-1,3-butadienes (Table 17) do not appear to agree with these conclusions.

Fulvene is well known to exhibit a diversity of regioselectivities in cycloaddition reactions.[132] Recognizing the potential suitability of 6,6-dimethylfulvene to differentiate electrophilic and nucleophilic carbenes, Moss et al.[135] found that CCl_2 reacted preferentially at the endocyclic double bond to produce, following rearrangement, **26** and **27** (Equation 25).[135]

$$\text{(25)}$$

In contrast, both MeOCCl and $(MeO)_2C$ underwent addition solely at the exocyclic double bond to produce **28**

(Z=Cl, OMe), although low yields characterized these reactions. Differential frontier orbital energy terms indicate dominant CCl_2-LUMO/dimethylfulvene-HOMO interaction (electrophilic addition) and dominant $(MeO)_2C$-HOMO/dimethylfulvene-LUMO interaction (nucleophilic addition). With ambiphilic MeOCCl, differential energies of the frontier molecular orbital interactions are nearly identical and, in this case, the mode of addition appears to be determined by the largest possible overlap during addition.

F. 1,2- vs. 1,4-Cycloaddition

The normal mode of carbene cycloaddition to dienes and polyenes is 1,2. Despite numerous attempts to achieve 1,4-addition, very few examples of this reaction mode have been reported[136-139] and, in each case, substantial preference is exhibited for 1,2-addition (e.g., Equation 26[136]). Hoffmann ascribed this preference to excessive closed shell interaction which is

Table 18
PRODUCT SELECTIVITY IN THE ADDITION OF
DIFLUOROCARBENE TO 2-SUBSTITUTED
NORBORNADIENES

29, Z=	exo-1,2 adducts[a]		endo-homo-1,4 adducts[a]		exo-1,2/homo-1,4 ratio[a]	IP (eV)[b]
	30	**31**	**32**	**33**		
OMe	89		11		8.1	8.05
OSiMe$_3$	75	6	6		4.2[c]	8.06
Ph	75		25		3.0	(8.51)[d]
Cl	62	4	34		1.9	8.77
H	33	33	34		1.9	8.69
COOEt	44	12	44		1.3	—
COOMe	38	15	47		1.1	8.92
CN	35	23	30	12	1.4	9.26

[a]   Reference 147, yields are normalized to 100%.
[b]   Reference 148.
[c]   Three other products (13%) have eluded characterization.
[d]   Average of first and second ionization potentials of 2-phenylbicyclo[2.2.1]hepta-2,5-diene.

$$99 \quad : \quad 1 \qquad (26)$$

insensitive to substituent effects.[140] Consequently, reports of 1,4-addition must be viewed skeptically, because their occurrence could be due to addition of a triplet carbene[138] to an intermediate ylide,[141] or to the vinylcyclopropane-cyclopentene rearrangement.[136]

In a series of publications extending over a 10-year period, Jefford and co-workers[142-148] have thoroughly described the novel homo-1,4 carbene additions to norbornadiene and substituted norbornadienes. Four products, two from *exo*-1,2-addition (**30** and **31**) and two from *endo*-homo-1,4-addition (**32** and **33**) are observed in reactions of CF$_2$ with 2-substituted norbornadienes (Equation 27), and the nature of the substituent Z has a substantial effect on reaction selectivity (Table 18).[145] Although the exo products are derived solely from 1,2-

$$(27)$$

addition, carbene attack from the endo side can result in products from both 1,2- and homo-1,4 addition. Competition between homo-1,4 and endo-1,2 addition is temperature dependent[144] but, even at 25°C, homo-1,4 addition is favored by more than 8 to 1. Selectivities from reactions of norbornadiene with CF_2, CClF,[144] CCl_2, and CBr_2[146] have been reported, as have additions to other substituted norbornadienes (e.g., Equation 28).[144,147]

$$(28)$$

A relatively complete explanation of these selectivities has emerged from molecular orbital calculations and spectroscopic determinations for 2-substituted norbornadienes.[148] Interaction of the carbene LUMO with the alkene HOMO constitutes the dominant frontier molecular orbital interaction for difluorocarbene.[78] The preferred site of attack for electrophilic carbenes such as CF_2 is the site of larger HOMO coefficients so that the HOMO-LUMO overlap is maximized.[132] Molecular orbital calculations at the STO-G3 level show that all norbornadiene 2-substituents cause the HOMO to be concentrated on the substituted double bond cations, which explains the observed preference for **30** and **31**.[148] The strong preference for **32** over **33** is also explicable in terms of the HOMO coefficients: interaction of the carbene LUMO occurs most strongly at C-3, which is the site of the largest HOMO coefficient. Finally, changes in the exo-1,2/homo-1,4 ratio as the 2-substituent is varied correspond with norbornadiene ionization potentials (Table 18), and exo-1,2 addition is retarded to a greater extent than is homo-1,4-addition by the decrease in the ''electron-richness'' of the norbornadiene.

IV. REARRANGEMENTS IN CARBENES

A. 1,2-Hydrogen Migrations

The most common rearrangement reaction of alkylcarbenes is the 1,2-shift of a hydrogen to the carbene center which generates an alkene.[149] This is generally recognized as a facile process. Few examples documented the competitiveness of intermolecular carbene reactions with the 1,2-hydrogen shift prior to 1970.[150,151] However, reports of cyclopropanation of olefins by methylchlorocarbene[53] and related diazirine-derived alkylhalocarbenes[3-7] demonstrated that the intermolecular reactions could be investigated with carbenes which undergo hydrogen migration.

Hoffmann et al.[71] presented semiempirical molecular orbital predictions as early as 1968. They claimed that the hydrogen which migrates to the electron-deficient carbon is aligned with the LUMO (p-orbital) of the singlet carbene (**34**).[71] Selectivity in the formation of olefins is reported to be dependent on both electronic and steric factors,[152] but increasing evidence points to stereoelectronic control.[121,153,154]

34

Table 19
DEPENDENCE OF ALKENE ISOMER RATIO FROM THERMAL DECOMPOSITION OF 3-HALO-3-ETHYLDIAZIRINES ON REACTION SOLVENT[161]

Solvent	Temp. (°C)	*(Z)*-ClCH=CHCH₃ / *(E)*-ClCH=CHCH₃	Solvent	Temp. (°C)	*(Z)*-BrCH=CHCH₃ / *(E)*-BrCH=CHCH₃
Cyclohexene	115	4.5	1-Dodecene	100	3.9
Anisole	95	4.2	Anisole	100	3.6
DMSO	100	2.8	DMSO	100	2.7
HMPA	100	2.0	HMPA	100	1.8

Press and Shechter have shown, for example, that cyclohexylidenes exhibit considerable axial stereochemical preference for migration (Equation 29).[153]

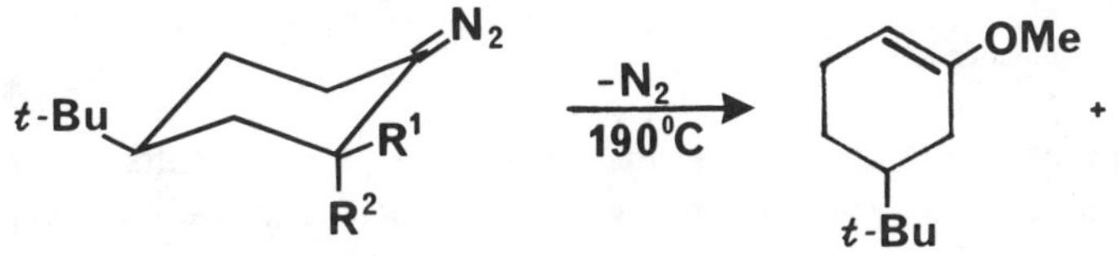

$$\textbf{35a} \ (R^1=OMe, R^2=H) \quad 95.3 \quad 4.7 \quad 0.0 \quad 0.0$$

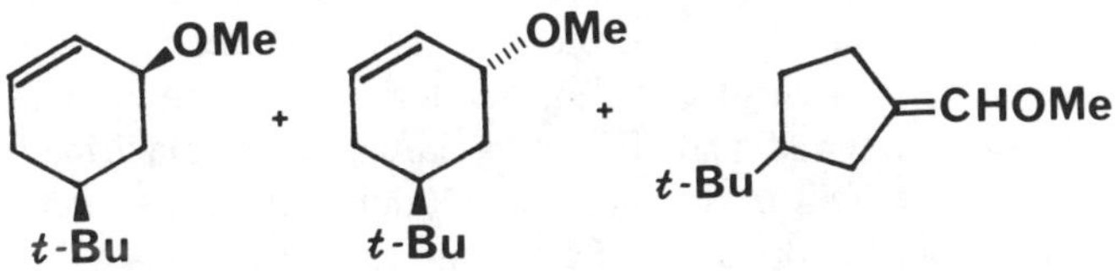

$$\textbf{35b} \ (R^1=H, R^2=OMe) \quad 31.4 \quad 0.0 \quad 54.5 \quad 14.1 \tag{29}$$

This communication afforded a rationale for an earlier report of low selectivity (migratory ratio, H(axial)/H(equatorial) = 1.50) for 1,2-hydrogen migrations with 4-*tert*-butyl-2,2-dimethylcyclohexylidene[156] which was in conflict with prior theoretical predictions.[157,158]

Liu has recently obtained the activation energy for 1,2-hydrogen migration in benzylchlorocarbene.[121] The value of 6.4 kcal/mol, obtained through competitive cyclopropanation with tetramethylethylene, is the first experimental report of an energy barrier for the 1,2-hydrogen shift. Although the estimated error is ±2 kcal/mol, this activation energy provides a valuable beginning for evaluation of factors that control this intramolecular rearrangement. More recently, he has found the activation energy for 1,2-hydrogen migration in benzylbromocarbene to be 4.7 kcal/mol.[155]

Arylalkylcarbenes produce *trans*-alkenes preferentially,[159,160] and dominance of the *(Z)*-olefin isomer has been reported for alkylhalocarbenes,[161] but not for benzylchlorocarbene.[116] As revealed in Section III.C and in Tables 14 and 15, the *(E)*/*(Z)*-isomer ratio for β-chlorostyrene from 1,2-hydrogen migration in PhCH₂CCl is dependent on both the structure and amount of olefin employed in competitive experiments.[117-119] A similar dependence of olefin isomer ratio on solvent has been reported for the thermal decomposition 3-bromo-3-ethyldiazirine and of 3-chloro-3-ethyldiazirine (Table 19).[161] Although originally interpreted

Table 20
PRODUCT COMPOSITION FROM PHOTOLYTIC AND THERMAL DECOMPOSITION OF CYCLOALKANESPIRODIAZIRINES[162]

40, n =	Condition (°C)	% Yield[a]				
		41	42	43	44	45
5	160	100.0	0.0			0.0
	hν, 0	99.6	0.3			0.1
6	160	100.0	0.0			0.0
	hν, 0	97.3	2.5	Trace		0.2
7	160	84.7	15.2	0.1		Trace
	hν, 0	82.8	15.1	1.8		0.3
8	160	47.0	9.0	0.0	44.0	Trace
	hν, 0	44.9	11.3	0.0	43.4	0.4
Diazocyclooctane	hν, 0	65.3	5.4	0.0	29.3	

[a]   Normalized to 100%.

as due to preferential solvation around the C–Cl dipole, these results could also arise from a carbene-solvent complex (ylide) which would be expected to cause a decreased preference for the less stable olefin isomer.

Intramolecular competition between 1,2-hydrogen migration, 1,2-alkyl migration, and C–H insertion of cycloalkylidenes has been reported.[162] Thermal and photochemical decomposition of cycloaklanespirodiazirines generated a mixture of products (Equation 30) whose composition varies with ring size (Table 20). Isomerization of the diazirine to the diazo compound accounts for between 30 and 60% of the initial fate of diazirine in photochemical reactions. The use of acetic acid in the reaction solution to trap the intermediate carbenes did not substantially inhibit the formation of C–H insertion products **42—44**, indicating that intermolecular trapping of secondary alkylcarbenes does not compete with intramolecular processes under these conditions.

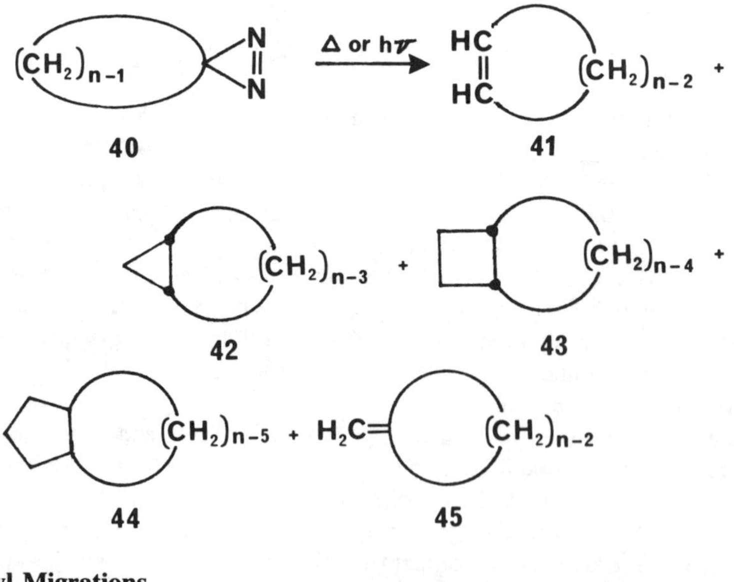

$$(30)$$

B. 1,2-Alkyl Migrations

Although not normally as facile nor as widely investigated as 1,2-hydrogen migrations, alkyl migrations can be preferred processes if constraints such as ring strain are imposed on the reacting system. Cyclopropyl- and cyclobutylcarbenes, for example, undergo ring expansion in preference to hydrogen migrations (e.g., Equation 31) or to phenyl migration.[70,163-166]

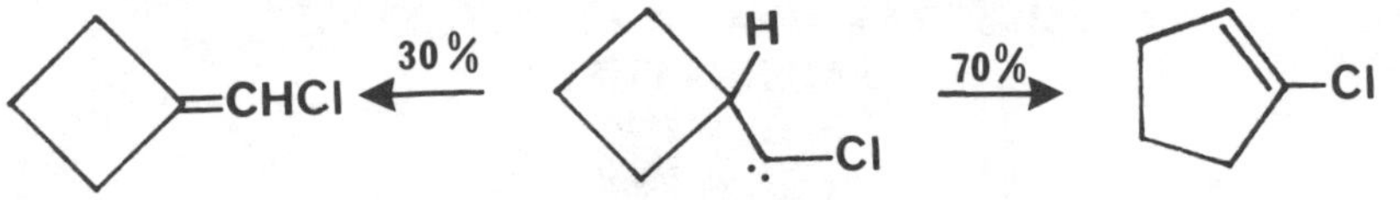

$$\text{(31)}$$

Recently, dicyclopropylcarbene has been generated by thermolysis of 5,5-dicyclopropyl-2-methoxy-2-methyl-Δ^3-1,3,4-oxa-diazoline (**46**) at 80°C and, in the absence of solvents prone to serve as atom donors to **47**, rearranges to 1-cyclopropyl-1-cyclobutene (Equation 32).[165] Selective alkyl shifts in bicyclic carbenes have also been reported.[167,168]

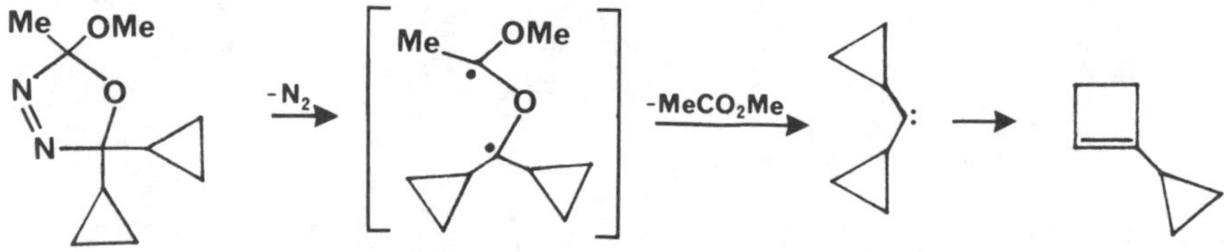

$$\text{(32)}$$

Sheridan and Kasselmayer have recently generated methoxychlorocarbene photolytically from the corresponding diazirine at 10 K in an argon matrix.[169] In contrast to thermolysis (25°C) results,[82,89] MeOCCl undergoes 1,2-O → C migration (Equation 33). This novel rearrangement occurs through the electronically excited carbene and is estimated to be exothermic by about 30 kcal/mol. The utilization of bond migrations in carbene intermediates for the preparation of high energy compounds is becoming increasingly important.[170-172]

$$\mathrm{Me\ddot{O}\ddot{C}Cl \xrightarrow{h\nu} MeCOCl} \qquad \text{(33)}$$

Although 1,2-alkyl or 1,2-hydrogen migrations often occur to the near exclusion of intermolecular reactions, Moss and Wetter have shown that at least with cyclopropylphenylcarbene,[164] intermolecular addition can become the major reaction process at low temperatures (Table 21). Irradiation of cyclopropylphenyldiazomethane at 25°C in the presence of isobutylene results in four volatile products (Equation 34) of which that from intermolecular cyclopropanation is barely noticeable. Even at −77°C, however, the yield of **49** exceeds that of **48**. The temperature effect on carbene reactions offers considerable flexibility in viewing intermolecular processes of alkylcarbenes previously thought to have been inaccessible.[162]

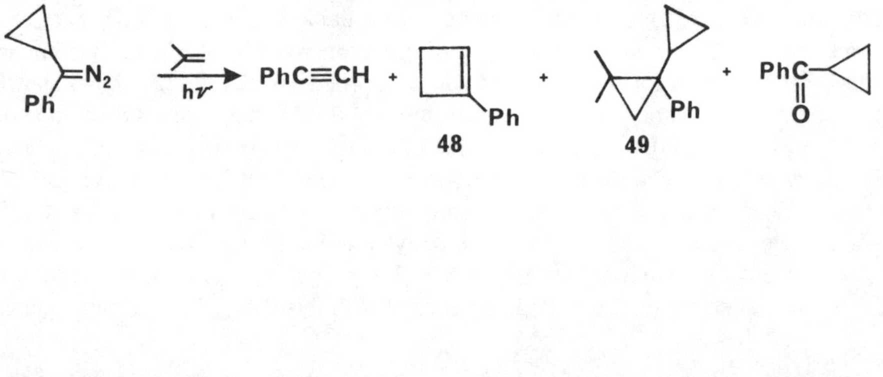

$$\text{(34)}$$

Table 21

**PRODUCT DISTRIBUTION FROM THE
PHOTOLYTIC DECOMPOSITION OF
CYCLOPROPYLPHENYLDIAZOMETHANE IN
ISOBUTYLENE AS A FUNCTION OF
TEMPERATURE[164]**

Temp. (°C)	% Yield				
	PhC≡CH	48	49	PhCOCp[a]	48/49
25	16.0	63	2.8	2.5	22.0
0	16.0	57	4.8	2.7	12.0
−18	16.0	52	7.4	3.3	7.1
−40	12.0	52	15.0	4.9	3.4
−77	3.9	27	34.0	1.9	0.76
−95	7.0	18	39.0	1.7	0.46
−128	4.4	10	42.0	3.3	0.24

[a]   Cyclopropyl phenyl ketone.

V. INSERTION REACTIONS

A. O–H Bond Insertion

A considerable body of information has recently become available concerning the reactions of carbenes with alcohol and acids.[173-180] Thermal and/or photolytic decompositions of diazirines in the presence of carboxylic acids were initially employed to monitor diazirine rearrangements to diazoalkanes.[159,181] When such a rearrangement fails to occur, as in decompositions of alkyl- and arylchlorocarbenes, the insertion reaction can be studied directly. Three distinct reaction mechanisms can be envisioned for the insertion process: (1) initial protonation followed by anion association, (2) direct insertion, and (3) anion addition followed by protonation. In their recent communication on the reaction of hydrogen chloride with benzylchlorocarbene, Liu and Subramanian proposed that insertion involves initial protonation[177] and, as evidence for this pathway, they cite the production of 1-phenyl-2,2-dichloroethane and β-chlorostyrene, variation in the $(E)/(Z)$-β-chlorostyrene isomer ratio with hydrogen chloride concentration, and deuteration of both reaction products in reactions with DCl. Absolute rate constants for the reactions of arylchlorocarbenes with acetic and trifluoroacetic acids have also been reported.[174] The O–H bond insertion by phenylchlorocarbene into acetic acid is more than one order of magnitude faster than addition to 2,3-dimethyl-2-butene (Table 22).

The insertion reactions of carbenes into the O–H bonds of alcohols have received considerable attention.[173,176,178-180] Chlorocarbenes have been shown to have a strong discrimination in their reactivity toward the monomers and oligomers of representative alcohols.[173] The methanol monomer, for example, is about two orders of magnitude less reactive than its oligomers toward O–H bond insertion by p-methoxyphenylchlorocarbene. A comparison of absolute rate constants for insertion and addition processes (Table 22) shows that O–H bond insertion into methanol oligomers is only slightly slower than into acetic acid and that both are near diffusion controlled. Reactions of arylchlorocarbenes with acetic acid or methanol are slower in acetonitrile than in isooctane, whereas rate constants for addition to tetramethylethylene do not show a pronounced solvent effect, presumably because of hydrogen bonding to acetonitrile.[173] Indeed, the activation energy for O–H bond insertion by ArCCl to methanol in acetonitrile is -4.7 ± 0.3 kcal/mol, which is in the range of the hydrogen bond strength for methanol.[173] In agreement with this value, Liu and Subramanian[176]

Table 22
ABSOLUTE RATE CONSTANTS FOR REACTIONS OF ARYLCHLOROCARBENES WITH ALCOHOLS, ACIDS, AND ALKENES[a]

Reactant	$k_{isooctane}$, M/sec		k_{CH_3CN}, M/sec		Ref.
	PhCCl	AnCCl[b]	PhCCl	AnCCl[b]	
$(MeOH)_{oligomer}$	2.9×10^9	4.3×10^9			173
$(MeOH)_{monomer}$		2×10^7		6.5×10^6	173
t-BuOH		2.5×10^6			173
CH_3COOH	3.1×10^9	5.1×10^9	1.8×10^9	2.2×10^9	174
CF_3COOH			2.4×10^9		174
$Me_2C=CMe_2$	$2.8 \times 10^{8\ c}$	1.4×10^7	2.0×10^8	2.0×10^8	105, 174
$Me_2C=CHMe$	$1.3 \times 10^{8\ c}$	7.7×10^6			105

[a]   Reactions performed at 300 K.
[b]   An = p-MeOC$_6$H$_4$.
[c]   Reactions performed at 296 K.

have established the activation energy for O–H bond insertion to be -4.5 kcal/mol for the trapping of benzylchlorocarbene by methanol, and they found that the insertion reaction is second order in methanol. *tert*-Butyl alcohol reacts as a monomer with ArCCl, and its activation energy for insertion is $+3.2 \pm 0.6$ kcal/mol.[173]

A detailed mechanism for termolecular O–H bond insertion has been proposed (Scheme 7).[180]

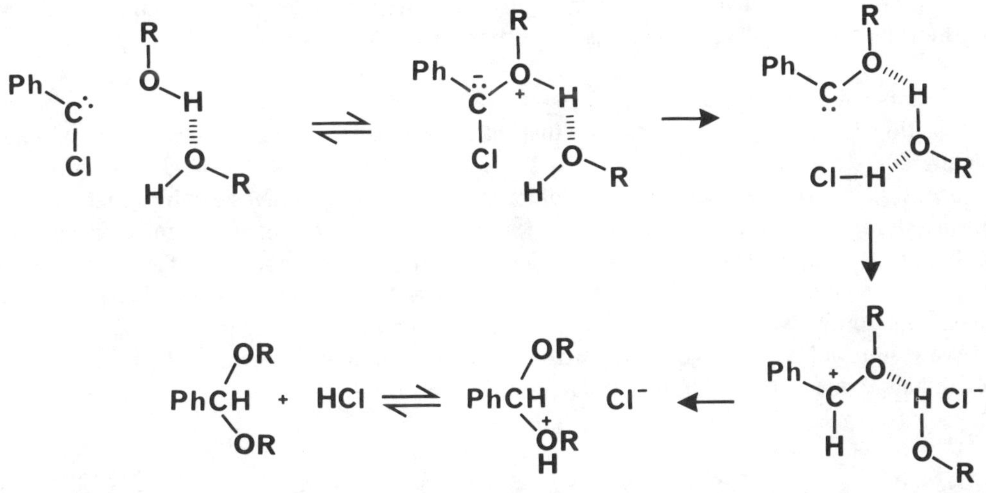

SCHEME 7.

According to this scheme, the initial product from O–H bond insertion is not that from direct insertion, as was originally suggested,[173] but an ylide intermediate, as was recently proposed by Warner and Chu.[178] Chloride loss from the ylide intermediate forms an ambiphilic carbene that remains hydrogen bonded to the alcohol. Although this proposal is speculative, it is consistent with available data. The inference that chlorocarbenes are susceptible to substitution is worthy of further investigation.

Warner and Chu[178] have investigated the mechanism of O–H bond insertion by carbenes.

They considered three general pathways for product formation from reactions of aliphatic carbenes:[182] (1) initial protonation to give a carbocation (Equation 35), (2) possibly reversible ylide formation (Equation 36), and (3) direct, three-center O–H bond insertion (Equation 37).

$$R_2C: + MeOH \rightarrow R_2\overset{+}{C}H + MeO^- \rightarrow R_2CHOMe \tag{35}$$

$$R_2C: + MeOH \rightleftharpoons R_2\overset{+}{\underset{H}{\overset{..}{C}-O}}Me \rightarrow R_2CHOMe \tag{36}$$

$$R_2C: + MeOH \rightarrow R_2CHOMe \tag{37}$$

Based on temperature-dependent isotope effects and product ratios, they conclude that initial protonation of **50** and reversible ylide formation with **51** account for their data.

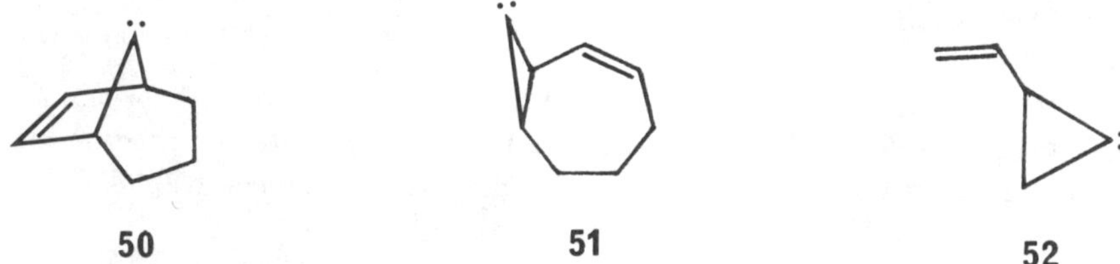

They considered three general.

50   **51**   **52**

Kirmse et al.[183] had previously explained the rearrangement of **52** to cyclopentenyl derivatives in the presence of methanol by the initial protonation mechanism. A clear rationale for the apparent mechanistic diversity in these insertion reactions is as yet lacking.

B. C–H Bond Insertion

Equation 30 and Table 20 explained that intramolecular C–H bond insertion by carbenes preferentially forms five-membered rings.[162] Insertion into a β–C–H bond is surprisingly competitive in this case, even though cyclopropane formation involves a substantial increase in ring strain. In contrast, formation of a four-membered ring is of minor importance. The earlier literature on C–H bond insertions leads to the same conclusions: ring formation occurs with the preference 5 > 6 > 3 >> 4.[8] In some cases (e.g., **40**, $n = 8$), five-membered ring formation is even favored over 1,2-hydrogen migration.

Crow and McNab have generated *ortho*-substituted arylalkylcarbenes **53—56** in the gas phase and explored their intramolecular insertion reactions.[184-188]

53   **54**   **55**   **56**

Their results confirm ring size preference obtained from studies of carbenes in solution, and they find that the presence of a heteroatom at the 4-position increases the preference for formation of the five-membered ring. In their investigations of **56**, they reported a 1,2-O → C migration of the rearranged carbene (Equation 38)[186]

$$56 \longrightarrow \quad \text{[structure: 2-CD-oxy-6-CHD}_2\text{ benzene]} \quad \longrightarrow \quad \text{[structure: 2-(C(=O)-D)-6-CHD}_2\text{ benzene]} \tag{38}$$

which was subsequently observed with methoxychlorocarbene (Equation 33) in an argon matrix.[169]

Tomioka and Izawa reported several years ago that photolysis of phenyldiazomethane in alcohols produced products from both C–H and O–H bond insertions (e.g., Equation 39).[189]

$$PhCH=N_2 \xrightarrow[CH_3OH]{h\nu} PhCH_2OCH_3 + PhCH_3 + (PhCH_2)_2 + PhCH_2CH_2OH \tag{39}$$

Decreasing the temperature from 0 to $-196°C$ caused a dramatic shift from O–H bond insertion to C–H bond insertion. The yields of toluene and bibenzyl were relatively invariant with temperature, and C–H bond insertion was attributed to the triplet carbene. This phenomenon has been explained by hydrogen atom tunneling[190] and, recently, Oikawa and Tsuda reported *ab initio* calculations demonstrating that the extraordinary temperature dependence originates from a stable intermediate formed between the carbene and alcohol.[179] They concluded that the contribution from the triplet carbene is not important.

A number of unanswered questions remain, not the least of which is the extent of stereoelectronic control in bond insertion processes. Few diazirines have been employed for evaluations of C–H bond insertions,[151,162] and little is known about the susceptibility of halocarbenes toward intramolecular bond insertion.

NOTE ADDED IN PROOF

Recent results by Doyle and co-workers[97] have clarified the mechanism of chlorophenylcarbene addition to diethyl maleate (Section III.B). The ratio of cyclopropane products (Equation 15) having the *cis*-dicarbethoxy stereochemistry to that having the transdicarbethoxy stereochemistry is lower in reactions of 3-chloro-3-*p*-nitrophenyldiazirine than in reactions of either 3-chloro-3-phenyldiazirine or the corresponding *m*-nitrophenyl derivative with diethyl maleate. In addition, diethyl maleate is converted to diethyl fumarate during the course of these reactions. These results are reported to be consistent with the formation of a carbene-alkene dipolar adduct whose lifetime is sufficiently long so that bond rotation can occur. A similar intermediate has been proposed by Liu to explain alkene-dependent 1,2-hydrogen migrations of benzylhalocarbenes,[121,155] and the general existence of intermediates in cyclopropanation reactions of singlet carbenes has been suggested by Giese.[191] Low temperature studies of the photochemical decomposition of 3-chloro-3-phenyldiazirine in the presence of diethyl maleate have shown that the apparent isomerization observed at or above room temperature is diminished until at $-196°C$ only the *cis*-dicarbethoxycyclopropane product is formed, which is also consistent with a carbene-alkene dipolar adduct. The isomerization of *trans*-cyclooctene to *cis* cyclobethene during the addition of CBr_2 (Equation 20)[111,112] is complimentary to results obtained with diethyl maleate.

REFERENCES

1. **Meier, H.,** Diazirine-diazoalkane interconversions, in *Chemistry of Diazirines,* Vol. 2, Liu, M. T. H., Ed., CRC Press, Boca Raton, Fla., 1987, chap. 6.
2. **Liu, M. T. H. and Stevens, I. D. R.,** Thermolysis and photolysis of diazirines, in *Chemistry of Diazirines,* Vol. 1, Liu, M. T. H., Ed., CRC Press, Boca Raton, Fla., 1987, chap. 5.
3. **Moss, R. A. and Jones, M., Jr.,** Carbenes, in *Reactive Intermediates,* Vol. 1, Jones, M., Jr. and Moss, R. A., Eds., Wiley-Interscience, New York, 1978.
4. **Moss, R. A. and Jones, M., Jr.,** Carbenes, in *Reactive Intermediates,* Vol. 2, Jones, M., Jr. and Moss, R. A., Eds., Wiley-Interscience, New York, 1981.
5. **Moss, R. A. and Jones, M., Jr.,** Carbenes, in *Reactive Intermediates,* Vol. 3, Jones, M., Jr. and Moss, R. A., Eds., Wiley-Interscience, New York, 1985.
6. **Jones, M., Jr. and Moss, R. A., Eds.,** *Carbenes,* Vol. 1, Wiley-Interscience, New York, 1973.
7. **Jones, M., Jr. and Moss, R. A., Eds.,** *Carbenes,* Vol. 2, Wiley-Interscience, New York, 1975.
8. **Kirmse, W.,** *Carbene Chemistry,* 2nd ed., Academic Press, New York, 1971.
9. **Rees, C. W. and Gilchrist, T. L.,** *Carbenes, Nitrenes, and Arynes,* Nelson, London, 1969.
10. **Hine, J.,** *Divalent Carbon,* Ronald Press, New York, 1964.
11. **Stang, P. J.,** Recent developments in unsaturated carbenes and related chemistry, *Acc. Chem. Res.,* 15, 348, 1982.
12. **Stang, P. J.,** Unsaturated carbenes, *Chem. Rev.,* 78, 383, 1978.
13. **Jones, W. M.,** Carbene-carbene rearrangements in solution, *Acc. Chem. Res.,* 10, 353, 1977.
14. **Moss, R. A.,** Carbenic selectivity in cyclopropanation reactions, *Acc. Chem. Res.,* 13, 58, 1980.
15. **Moss, R. A. and Pilkiewicz, F. G.,** Crown ethers in carbene chemistry. The generation of free phenyl-halocarbenes, *J. Am. Chem. Soc.,* 96, 5632, 1974.
16. **Moss, R. A. and Gerstl, R.,** Stereoselectivity of carbene intermediates. II. Phenylbromocarbene, *Tetrahedron,* 22, 2637, 1966.
17. **Moss, R. A., Joyce, M. A., and Pilkiewicz, F. G.,** Carbenes and carbenoids — crown ether and carbenic selectivity experiments with phenylfluorocarbene and chloromethylthiocarbene, *Tetrahedron Lett.,* 2425, 1975.
18. **Stang, P. J. and Mangum, M. G.,** Unsaturated carbenes from primary vinyl triflates. V. The nature of vinylidene carbene intermediates, *J. Am. Chem. Soc.,* 97, 6478, 1975.
19. **Moss, R. A., Joyce, M. A., and Huselton, J. K.,** The olefinic selectivity of dibromocarbene, *Tetrahedron Lett.,* 4621, 1975.
20. **Kirmse, W. and Von Wedel, B.,** Carbenes from geminal dihalides, *Annalen,* 666, 1, 1963.
21. **Friedman, L. and Shechter, H.,** Carbenoid and cationoid decomposition of diazo hydrocarbons derived from tosylhydrazones, *J. Am. Chem. Soc.,* 81, 5512, 1959.
22. **Mansorr, A. M. and Stevens, I. D. R.,** Hot radical effect in carbene reactions, *Tetrahedron Lett.,* 1733, 1966.
23. **Frey, H. M. and Scaplehorn, A. W.,** Thermal decomposition of the diazirines. II. 3,3-Tetramethylene-diazirine, 3,3-pentamethylenediazirine, and 3,3-diethyldiazirine, *J. Chem. Soc. Ser. A,* 968, 1966.
24. **Frey, H. M. and Stevens, I. D. R.,** The photolysis of 3-*t*-butyldiazirine, *J. Chem. Soc.,* 3101, 1965.
25. **Kirmse, W. and Horn, K.,** Comparison of catalylic, thermal, and photolytic decomposition of diazoalkanes, *Chem. Ber.,* 100, 2698, 1967.
26. **More O'Ferrall, R. A.,** The reactions of aliphatic diazo compounds with acids, *Adv. Phys. Org. Chem.,* 5, 31, 1967.
27. **Hegarty, A. F.,** Kinetics and mechanisms of reactions involving diazonium and diazo groups, in *The Chemistry of Diazonium and Diazo Groups,* Part 2, Patai, S., Ed., John Wiley & Sons, New York, 1978, chap. 12.
28. **Schmitz, E.,** Synthesis and reactions of diazirines, in *Chemistry of Diazirines,* Liu, M. T. H., Ed., CRC Press, Boca Raton, Fla., 1987, chap. 3.
29. **Oberberger, C. G. and Anselme, J.-P.,** The thermal and photolytic decomposition of 1-pentyldiazoethane, *J. Org. Chem.,* 29, 1188, 1964.
30. **Yates, P., Farnum, D. G., and Wiley, D. H.,** Aliphatic diazo compounds. VII. The reaction α-diazo ketones with diazomethane, *Tetrahedron,* 18, 881, 1962.
31. **Bethell, D., Newall, A. R., Stevens, G., and Wittaker, D.,** Intermediates in the decomposition of aliphatic diazo compounds. VII. Mechanisms for formation of benzophenone azine and diphenylmethanol in the thermal decomposition of diphenyldiazomethane, *J. Chem. Soc. Ser. B,* 749, 1969.
32. **Padwa, A. and Eastman, D.,** Reactions of phenylchlorodiazirine with nucleophiles and substituted acetylenes, *J. Org. Chem.,* 34, 2728, 1969.
33. **Gale, D. M., Middleton, W. J., and Krespan, C. G.,** Perfluorodiazo compounds, *J. Am. Chem. Soc.,* 88, 3617, 1966.

34. **Liu, M. T. H. and Ramakrishnan, K.,** Mechanism for the formation of azine in the decomposition of diazo-alkanes and diazirines, *Tetrahedron Lett.,* 3139, 1977.
35. **Doyle, M. P., Devia, A. H., Bassett, K. E., Terpstra, J. W., and Mahapatro, S. N.,** Unsymmetrical alkenes by carbene coupling from diazirine decomposition in the presence of diazo compounds, *J. Org. Chem.,* in press.
36. **Huisgen, R.** 1,3-Dipolar cycloadditions: past and future, *Angew. Chem. Int. Ed. Engl.,* 2, 565, 1963.
37. **Huisgen, R., Koszinowski, J., Ohta, A., and Schiffer, R.,** Cycloadditions of diazoalkanes to 1-alkenes, *Angew. Chem. Int. Ed. Engl.,* 19, 202, 1980.
38. **Doyle, M. P., Colsman, M. R., and Dorow, R. L.,** Effective methods for the synthesis of 2-pyrazolines from diazocarbonyl compounds, *J. Heterocyc. Chem.,* 20, 943, 1983.
39. **Doyle, M. P., Dorow, R. L., and Tamblyn, W. H.,** Cyclopropanation of α,β-unsaturated carbonyl compounds and nitriles with diazo compounds. The nature of the involvement of transition metal promoters, *J. Org. Chem.,* 47, 4059, 1982.
40. **Tomioka, H., Ohno, K., Izawa, Y., Moss, R. A., and Munjal, R. C.,** The philicity of a triplit carbene: additions of diphenylcarbene to styrene substrates, *Tetrahedron Lett.,* 25, 5415, 1984.
41. **Moss, R. A.,** Exchange reactions of halodiazirines, in *Chemistry of Diazirines,* Liu, M. T. H., Ed., CRC Press, Boca Raton, Fla., 1987, chap. 4.
42. **Heine, H. W.,** Diaziridines, 3H-diazirines, diaziridinones, and diaziridinimines, in *Small Ring Heterocycles,* Part 2, Hassner, A., Ed., John Wiley & Sons, New York, 1983, chap. 4.
43. **Skell, P. S. and Garner, A. Y.,** Reactions of bivalent carbon compounds. Reactivities in olefin-dibromocarbene reactions, *J. Am. Chem. Soc.,* 78, 5430, 1956.
44. **Skell, P. S. and Garner, A. Y.,** The stereochemistry of carbene-olefin reactions. Reactions of dibromocarbene with the *cis-* and *trans*-2-butenes, *J. Am. Chem. Soc.,* 78, 3409, 1956.
45. **Skell, P. S. and Woodworth, R. C.,** Structure of carbene, CH_2, *J. Am. Chem. Soc.,* 78, 4496, 1956.
46. **Skell, P. S. and Cholod, M. S.,** Reactions of dichlorocarbene with olefins. Temperature dependence of relative reactivities, *J. Am. Chem. Soc.,* 91, 7131, 1969.
47. **Doering, W. V. E. and Henderson, W. A.,** Electron-seeking demands of dichlorocarbene in its additions to olefins, *J. Am. Chem. Soc.,* 80, 5274, 1958.
48. **Doering, W. V. E. and LaFlamme, P.,** The *cis*-addition of dibromocarbene and methylene to *cis-* and *trans*-2-butene, *J. Am. Chem. Soc.,* 78, 5447, 1956.
49. **Moss, R. A., Mallon, C. B., and Ho, C.-T.,** The correlation of carbenic reactivity, *J. Am. Chem. Soc.,* 99, 4105, 1977.
50. **Moss, R. A. and Mallon, C. B.,** The characterization of carbene selectivity. Applications of difluorocarbene, *J. Am. Chem. Soc.,* 97, 344, 1975.
51. **Moss, R. A., Whittle, J. R., and Friedenreich, P.,** Stereoselectivity of carbene intermediates. VI. Selectivity of phenylchlorocarbene, *J. Org. Chem.,* 34, 2220, 1969.
52. **Moss, R. A.,** The photolysis of phenylbromodiazirine and the generation of phenylbromocarbene, *Tetrahedron Lett.,* 4905, 1967.
53. **Moss, R. A. and Mamantov, A.,** Methylchlorocarbene, *J. Am. Chem. Soc.,* 92, 6951, 1970.
54. **Moss, R. A., Vezza, M., Guo, W., Munjal, R. C., Houk, K. N., and Rondan, N. G.,** Reactivity of cyclopropylchlorocarbene: an interactive experimental and theoretical analysis, *J. Am. Chem. Soc.,* 101, 5088, 1979.
55. **Ehrenson, S., Brownless, R. T. C., and Taft, R. W.,** A generalized treatment of substituent effects in the benzene series. A statistical analysis by the dual substituent parameter equation, *Prog. Phys. Org. Chem.,* 10, 1, 1973.
56. **Moss, R. A. and Pilkiewicz, F. G.,** *gem*-Chloro(methylthio)cyclopropanes from *gem*-chloro(methylthio)carbene, *Synthesis,* 209, 1973.
57. **Moss, R. A. and Przybyla, J. R.,** Stereoselectivity of carbene intermediates. V. Phenylfluorocarbene, *Tetrahedron,* 25, 647, 1969.
58. **Moss, R. A., Guo, W., Denney, D. Z., Houk, K. N., and Rondan, N. G.,** Selectivity of (trifluoromethyl)chlorocarbene, *J. Am. Chem. Soc.,* 103, 6164, 1981.
59. **Nishimura, J., Furukawa, J., Kawabata, N., and Kitayama, M.,** The relative reactivity of olefins in cycloaddition with zinc carbenoid, *Tetrahedron,* 27, 1799, 1971.
60. **Newman, M. S. and Patrick, T. B.,** The reaction of dimethylethylidenecarbene with olefins, *J. Am. Chem. Soc.,* 91, 6461, 1969.
61. **Seyferth, D., Mui, J. Y. P., and Damrauer, R.,** Halomethyl-metal compounds. XIX. Further studies of the aryl(bromodichloromethyl)mercury-olefin reaction, *J. Am. Chem. Soc.,* 90, 6182, 1968.
62. **Durr, H. and Werndorff, F.** Proof of the electrophilic character of cyclopentadienylidene, *Angew. Chem. Int. Ed. Engl.,* 13, 483, 1974.
63. **Stang, P. J. and Mangum, M. G.,** Unsaturated carbenes from primary vinyl triflates. V. The nature of vinylidene carbene intermediates, *J. Am. Chem. Soc.,* 97, 6478, 1975.

64. **Levitt, L. S., Levitt, B. W., and Parkanyi, C.,** Inductive effects on molecular ionization potentials. VI. Alkenes, *Tetrahedron,* 28, 3369, 1972.

65. **Freeman, F.,** Possible criteria for distinguishing between cyclic and acyclic activated complexes and among cyclic activated complexes in addition reactions, *Chem. Rev.,* 75, 439, 1975.

66. **Schoeller, W. W., Aktekin, N., and Friege, H.,** Olefin selectivities in the addition of carbenes, *Angew. Chem. Int. Ed. Engl.,* 21, 932, 1982.

67. **Giese, B., Lee, W.-B., and Stiehl, C.,** Steric effects in carbene cycloadditions, *Tetrahedron Lett.,* 24, 881, 1983.

68. **Giese, B. and Nuemann, C.,** Temperature dependence of polar and steric effects in CCl_2 cycloadditions, *Tetrahedron Lett.,* 23, 3557, 1982.

69. **Taft, R. W., Jr.,** Separation of polar, steric, and resonance effects in reactivity, in *Steric Effects in Organic Chemistry,* Newman, M. S., Ed., John Wiley & Sons, New York, 1956, 598.

70. **Moss, R. A. and Fantina, M. E.,** Elusive carbenes. Olefinic capture of cyclopropylchlorocarbene and related species, *J. Am. Chem. Soc.,* 100, 6788, 1978.

71. **Hoffmann, R., Zeiss, G. D., and Van Dine, G. W.,** The electronic structure of methylenes, *J. Am. Chem. Soc.,* 90, 1485, 1968.

72. **Hahn, R. C., Corbin, T. F., and Shechter, H.,** Electrical effects of cycloalkyl groups, *J. Am. Chem. Soc.,* 90, 3404, 1968.

73. **Giese, B.,** Basis and limitations of the reactivity-selectivity principle, *Angew. Chem. Int. Ed. Engl.,* 16, 125, 1977.

74. **Giese, B., Leu, W.-B., and Meister, J.,** Selectivity of dihalocarbenes in cycloaddition reactions, *Annalen,* 725, 1980.

75. **Giese, B. and Lee, W.-B.,** Temperature dependence of linear free energy relationship of carbene cycloaddition, *Chem. Ber.,* 114, 3306, 1981.

76. **Giese, B. and Lee, W.-B.,** Experimental evidence for a correlation between isoselective and isokinetic temperatures, *Tetrahedron Lett.,* 23, 3561, 1982.

77. **Giese, B. and Lee, W.-B.,** The significance of isoselective temperatures for the existence of linear free-energy relationships, *Angew. Chem. Int. Ed. Engl.,* 19, 835, 1980.

78. **Rondan, N. G., Houk, K. N., and Moss, R. A.,** Transition states and selectivities of singlet carbene cycloadditions, *J. Am. Chem. Soc.,* 102, 1770, 1979.

79. **Hoffmann, R. W., Eillienblum, W., and Dittrich, B.,** Addition of dimethoxycarbene to C-C-multiple bonds, *Chem. Ber.,* 107, 3395, 1974.

80. **Hoffmann, R. W. and Reiffen, M.,** The nucleophilic character of dimethoxycarbene, *Chem. Ber.,* 109, 2565, 1976.

81. **Reiffen, M. and Hoffmann, R. W.,** On the reaction of amide acetals with heterocumulenes, *Chem. Ber.,* 110, 37, 1977.

82. **Moss, R. A. and Shieh, W.-C.,** Methoxychlorocarbene, *Tetrahedron Lett.,* 1935, 1978.

83. **Moss, R. A., Fedorynski, M., and Shieh, W.-C.,** Unification of the carbenic selectivity spectrum. The ambiphilicity of methoxychlorocarbene, *J. Am. Chem. Soc.,* 101, 4736, 1979.

84. **Moss, R. A., Perez, L. A., Wlostowska, J., Guo, W., and Krogh-Jespersen, K.,** Phenoxychlorocarbene, a second ambiphile, *J. Org. Chem.,* 47, 4177, 1982.

85. **Moss, R. A., Guo, W., and Krogh-Jespersen, K.,** Reactivity of an ambiphilic carbene — a nonlinear Hammett correlation, *Tetrahedron Lett.,* 23, 15, 1982.

86. **Moss, R. A. and Perez, L. A.,** The philicity of phenoxychlorocarbene towards styrenes, *Tetrahedron Lett.,* 24, 2719, 1983.

87. **Moss, R. A. and Munjal, R. C.,** Differentiation of electrophilic and ambiphilic carbenes, *Tetrahedron Lett.,* 4721, 1979.

88. **Doyle, M. P., Terpstra, J. W., and Winter, C. H.,** Nucleophilic character of an electrophilic carbene. Synthesis of cyclopropanes by thermal decomposition of 3-chloro-3-phenyldiazirine, *Tetrahedron Lett.,* 25, 901, 1984.

89. **Smith, N. P. and Stevens, I. D. R.,** Methoxychlorocarbene, *Tetrahedron Lett.,* 1931, 1978.

90. **Smith, N. P. and Stevens, I. D. R.,** Formation and reactions of chloro-methoxy- and -(2-methylpropoxy)carbene, *J. Chem. Soc. Perkin Trans. II,* 1298, 1979.

91. **Majchrzak, M. W., Kotelko, A., and Lambert, J. B.,** Palladium (II) acetate, an efficient catalyst for cyclopropanation reactions with ethyl diazoacetate, *Synthesis,* 469, 1983.

92. **Apeloig, Y., Karni, M., and Rappoport, Z.,** The role of hyperconjugation in determining the stereochemistry of nucleophilic epoxidation and cyclopropanation of electrophilic olefins, *J. Am. Chem. Soc.,* 105, 2784, 1983.

93. **Friedrich, K., Jansen, U., and Kirmse, W.,** Oxygen ylides. I. Reactions of carbenes with oxetane, *Tetrahedron Lett.,* 26, 193, 1985.

94. **Kirmse, W. and Chiem, P. V.,** Oxygen ylides. II. Photochemical and rhodium-catalyzed reactions of diazomethane with (S)-2-methyloxetane, *Tetrahedron Lett.,* 26, 197, 1985.

95. **Bekhazi, M. and Warkintin, J.,** Reactions of carbenes with acetone. Reversible thermal formation and fragmentation of carbonyl ylides and their cycloaddition to acetone, *J. Am. Chem. Soc.*, 105, 1289, 1983.

96. **Doyle, M. P., Griffin, J. H., Chinn, M. S., and van Leusen, D.,** Rearrangements of ylides generated from reactions of diazo compounds with allyl acetals and thioketals by catalytic methods. Heteroatom acceleration of the [2,3] sigmatropic rearrangement, *J. Org. Chem.*, 49, 1917, 1984.

97. **Doyle, M. P., Loh, K.-L., Nishioka, L., McVickar, M., and Liu, M. T. H.,** Formation of a dipolar adduct in the reaction of arylchlorocarbenes with diethyl maleate, *Tetrahedron Lett.*, 27, 4395, 1986.

98. **Moriarity, R. M., Bailey, B. R., III, Prakash, O., and Prakash, I.,** Capture of electron-deficient species with aryl halides. New syntheses of hypervalent iodoniom ylides, *J. Am. Chem. Soc.*, 107, 1375, 1985.

99. **Hoffmann, R.,** Trimethylene and the addition of methylene to ethylene, *J. Am. Chem. Soc.*, 90, 1475, 1968.

100. **Houk, K. N., Rondan, N. G., and Mareda, J.,** Are π-complexes intermediates in halocarbene cycloadditions?, *J. Am. Chem. Soc.*, 106, 4291, 1984.

101. **Houk, K. M. and Rondan, N. G.,** Origin of negative activation energies and entropy control of halocarbene cycloadditions and related reactions, *J. Am. Chem. Soc.*, 106, 4293, 1984.

102. **Cox, D. P., Gould, I. R., Hacker, N. P., Moss, R. A., and Turro, N. J.,** Absolute rate constants for the additions of halophenylcarbenes to alkenes; a reactivity-selectivity relation, *Tetrahedron Lett.*, 24, 5313, 1983.

103. **Turro, N. J., Lehr, G. F., Butcher, J. A., Jr., Moss, R. A., and Guo, W.,** Temperature dependence of the cycloaddition of phenylchlorocarbene to alkenes. Observation of "negative activation energies", *J. Am. Chem. Soc.*, 104, 1754, 1982.

104. **Turro, N. J., Butcher, J. A., Moss, R. A., Guo, W., Munjal, R. C., and Fedorynski, M.,** Absolute rate constants for additions of phenylchlorocarbene to alkenes, *J. Am. Chem. Soc.*, 102, 7576, 1980.

105. **Moss, R. A., Perez, L. A., Turro, N. J., Gould, I. R., and Hacker, N. P.,** Hammett analysis of absolute carbene addition rate constants, *Tetrahedron Lett.*, 24, 685, 1983.

106. **Giese, B. and Meister, J.,** Temperature dependence of carbene selectivities: a new example of the isoselective relationship, *Angew. Chem. Int. Ed. Engl.*, 17, 595, 1978.

107. **Giese, B., Lee, W.-B., and Neumann, C.,** Evidence for intermediates in the cycloaddition of single carbenes, *Angew. Chem. Int. Ed. Engl.*, 21, 310, 1982.

108. **Hammett, L. P.,** *Physical Organic Chemistry*, 2nd ed., McGraw-Hill, New York, 1970, chap. 12.

109. **Yang, N. C. and Marolewski, T. A.,** The addition of halomethylene to 1,2-dimethylcyclobutene, a methylene-olefin reaction involving a novel rearrangement, *J. Am. Chem. Soc.*, 90, 5654, 1968.

110. **Gale, D. M., Middleton, W. J., and Krespan, C. G.,** Perfluorodiazo compounds, *J. Am. Chem. Soc.*, 88, 3617, 1966.

111. **Cope, A. C., Moore, W. R., Bach, R. D., and Winkler, H. J. S.,** Molecular asymmetry. IX. The partial resolution and asymmetric synthesis of 1,2-cyclononadiene, *J. Am. Chem. Soc.*, 92, 1243, 1970.

112. **Dehmlow, E. V. and Kramer, R.,** Stereochemical evidence supporting the intermediate mechanism of dihalocarbene addition, *Angew. Chem. Int. Ed. Engl.*, 23, 706, 1984.

113. **Liu, M. T. H. and Toriyama, K.,** Pyrolysis of 3-chloro-3-aryldiazirines, kinetics and mechanism, *Can. J. Chem.*, 50, 3009, 1972.

114. **Shimizu, N. and Nishida, S.,** A new substrate to investigate radical cycloadditions. II. Addition reactions of some representative carbenes to 1,1-dicyclopropylethylene, *J. Am. Chem. Soc.*, 96, 6451, 1974.

115. **Lambert, J. B., Larson, E. G., and Bosch, R. J.,** Nonstereospecific, nontriplet cyclopropanations with dibromomethylene, *Tetrahedron Lett.*, 24, 3799, 1983.

116. **Liu, M. T. H., Chishti, N. H., Tencer, M., Tomioka, H., and Izawa, Y.,** Reactive intermediates in the photolysis and thermolysis of 3-chloro-3-benzyldiazirine, *Tetrahedron*, 40, 887, 1984.

117. **Tomioka, H., Hayashi, N., Izawa, Y., and Liu, M. T. H.,** Photolysis of 3-chlorodiazirine in the presence of alkene intermediate in addition of chlorocarbene to alkene, *J. Am. Chem. Soc.*, 106, 454, 1984.

118. **Tomioka, H., Hayashi, N., Izawa, Y., and Liu, M. T. H.,** Photolysis of 3-chlorodiazirine in the presence of alkenes. Kinetic evidence for intervention of a carbene-alkenes: orientational factors in the intermolecular interception of chlorocarbene by alkenes, *J. Chem. Soc. Chem. Commun.*, 476, 1984.

119. **Liu, M. T. H. and Subramanian, R.,** Addition of an electrophilic carbene to an electron-deficient olefin. Kinetics of benzylchlorocarbene-diethyl fumarate reaction, *Tetrahedron Lett.*, 26, 3071, 1985.

120. **Warner, P. M.,** On carbene alkene complexes, *Tetrahedron Lett.*, 25, 4211, 1984.

121. **Liu, M. T. H.,** Energy barrier for 1,2-hydrogen migration in benzylchlorocarbene, *J. Chem. Soc. Chem. Commun.*, 982, 1985.

122. **Moss, R. A. and Munjal, R. C.,** Stereoselectivity of cyclopropylchlorocarbene, *Tetrahedron Lett.*, 21, 2037, 1980.

123. **Barlet, R., Le Goaller, R., and Gey, C.,** Méthylhalogénocyclopropanation des éthoxy-alcènes diversement alkylés. Identification stéréochemique des adduits et étude de la stéréosélectivité, *Can. J. Chem.*, 60, 1933, 1982.

124. **Blanco, L., Amice, P., and Conia, J.-M.,** Reaction of chloromethylcarbene with trimethylsilyl enol ethers; conversion of cycloalkanones into higher α-methylcycloalkenones, *Synthesis,* 289, 1981.

125. **Doyle, M. P. and McVickar, M.,** unpublished results.

126. **Doyle, M. P., Dorow, R. L., Tamblyn, W. H., and Buhro, W. E.,** Regioselectivity in catalytic cyclopropanation reactions, *Tetrahedron Lett.,* 23, 2261, 1982.

127. **Skattebøl, L.,** Chemistry of *gem*-dihalocyclopropanes. II. The reaction of dienes with dibromocarbene, *J. Org. Chem.,* 25, 2951, 1964.

128. **Kostikov, R. R. and Bespalov, A. Ya.,** Reaction of dichlorocarbene with 2-phenyl-1,3-butadiene, *Zh. Org. Khim.,* 6, 629, 1970.

129. **D'yakonov, I. A., Kornilova, T. A., and Nizovkina, B.,** Carbene reactions with diene, enyne, and diyne systems. IV. Reactions of dichlorocarbene with chloroprene, *Zh. Org. Khim.,* 3, 272, 1967.

130. **D'yakanov, I. A., Kostikov, R. R., and Aksenov, V. S.,** Reaction of carbenes with conjugated di- and polyene compounds. I. Reaction of dichloro- and dibromocarbenes with *trans*-1-phenyl-1,3-butadiene, *Zh. Org. Khim.,* 6, 1965, 1970.

131. **D'yakanov, I. A. and Kornilova, T. A.,** Reactions of dichlorocarbene with 1-chloro-1,3-butadiene, *Zh. Org. Khim.,* 5, 178, 1969.

132. **Houk, K. N.,** The frontier molecular orbital theory of cycloaddition reactions, *Acc. Chem. Res.,* 8, 361, 1975.

133. **Sauer, J.,** Diels-Alder reactions II: the reaction mechanism, *Angew. Chem. Int. Ed. Engl.,* 6, 17, 1967.

134. **Rondan, N. G. and Houk, K. N.** Transition structures for the reaction of difluorocarbene with propene, *Tetrahedron Lett.,* 25, 5965, 1984.

135. **Moss, R. A., Young, C. M., Perez, L. A., and Krogh-Jespersen, K.,** Carbenic philicity: 6,6-dimethylfulvene as an indicator substrate, *J. Am. Chem. Soc.,* 103, 2413, 1981.

136. **Turkenburg, L. A. M., de Wolf, W. H., and Bickelhaupt, F.,** 1,4-Addition of dichlorocarbene to 1,2-bismethylenecycloheptane, *Tetrahedron Lett.,* 23, 769, 1982.

137. **Anastassiow, A. G., Cellura, R. P., and Ciganek, E.,** 1,4-Addition of dicyanocarbene onto cyclooctatetraene, *Tetrahedron Lett.,* 5267, 1970.

138. **Franzen, V.,** 1,4-Addition of methylene to butadiene, *Chem. Ber.,* 95, 571, 1962.

139. **Dehmlow, E. V., Broda, W., and Schladerbeck, N. H.,** Dihalocarbene reactions of isobenzofurans, *Tetrahedron Lett.,* 25, 5509, 1984.

140. **Fujimoto, H. and Hoffmann, R.,** A molecular orbital study of the addition of singlet methylene to butadiene, *J. Phys. Chem.,* 78, 1167, 1974.

141. **Krageloh, K., Anderson, G. H., and Stang, P. J.,** Ylide vs. 1,4-cycloaddition in the interaction of an alkylidenecarbene with azoarenes and the formation of 2*H*-indazoles and tetrahydrotetrazoles, *J. Am. Chem. Soc.,* 106, 6015, 1984.

142. **Jefford, C. W., Kabengele, nT., Kovacs, J., and Burger, U.,** Additions of fluorocarbenes to norbornadiene by linear cheletropic reactions, *Tetrahedron Lett.,* 257, 1974.

143. **Jefford, C. W., Kabengele, nT., Kovacs, J., and Burger, U.,** The addition of fluorocarbenes to norbornadiene. A homo-1,4-addition, *Helv. Chim. Acta,* 57, 104, 1974.

144. **Jefford, C. W., Mareda, J., Gehret, J.-C. E., Kabengele, nT., Graham, W. D., and Burger, U.,** Cheletropic reactions of fluorocarbenes with norbornadienes, *J. Am. Chem. Soc.,* 98, 2585, 1976.

145. **Jefford, C. W. and Phan Thanh, H.,** Competing cheletropic reactions of difluorocarbene with various 2-substituted norbornadienes, *Tetrahedron Lett.,* 755, 1980.

146. **Jefford, C. W., de los Heros, V., Gehret, J. C. E., and Wipff, G.,** The role of entropic and electronic factors in controlling homo-1,4 addition of dihalocarbenes to norbornadiene-type molecules, *Tetrahedron Lett.,* 1629, 1980.

147. **Jefford, C. W. and Phan Thanh, H.,** Effect of substituents on the homo-1,4 addition of difluorocarbene to norbornadiene, *Tetrahedron Lett.,* 23, 391, 1982.

148. **Houk, K. N., Rondan, N. G., Paddon-Row, M. N., Jefford, C. W., Phan Thanh, H., Burrow, P. D., and Jordan, K. D.,** Ionization potentials, electron affinities, and molecular orbitals of 2-substituted norbornadienes. Theory of 1,2 and homo-1,4 carbene cycloaddition selectivities, *J. Am. Chem. Soc.,* 105, 5563, 1983.

149. **Schaeffer, H. F., III,** The 1,2-hydrogen shift: a common vehicle for the appearance of evanescent molecular species, *Acc. Chem. Res.,* 12, 288, 1979.

150. **Frey, H. M.,** Photolysis of the diazirines, *Adv. Photochem.,* 4, 225, 1966.

151. **Frey, H. M. and Liu, M. T. H.,** The thermal decomposition of the diazirines. IV. 3-Chloro-3-ethyldiazirine, 3-chloro-3-*n*-propyldiazirine, 3-chloro-3-isopropyl-diazirine, and 3-chloro-3-*t*-butyldiazirine, *J. Chem. Soc. Ser. A,* 1916, 1970.

152. **Yamamoto, Y. and Moritani, I.,** Stereochemistry in the 1,2-hydrogen migration to a divalent carbon, *Tetrahedron,* 26, 1235, 1970.

153. **Press, L. S. and Shechter, H.,** The case for stereoelectronic axial migration of hydrogen in nonrigid cyclohexylidenes in chair conformation, *J. Am. Chem. Soc.,* 101, 509, 1979.

154. **Liu, M. T. H. and Tencer, M.**, Electronic effect on 1,2-hydrogen shift in chlorocarbenes, *Tetrahedron Lett.*, 24, 5713, 1983.

155. **Liu, M. T. H. and Subramanian, R.**, Energy barrier for 1,2-hydrogen migration in benzylbromocarbene, *J. Phys. Chem.*, 90, 75, 1986.

156. **Seghers, L. and Shechter, H.**, The stereochemistry of decomposition of 1-diazo-2-phenyl-5-*t*-butylcyclohexanes, *Tetrahedron Lett.*, 1943, 1976.

157. **Altmann, J. A., Csizmadia, I. G., and Yates, K.**, An *ab initio* study of methylcarbene and the stereochemistry of its rearrangement of ethylene, *J. Am. Chem. Soc.*, 96, 4196, 1974.

158. **Altmass, J. A., Tee, O. S., and Yates, K.**, Application of the principle of least motion of organic reactions. IV. More complex molecular rearrangements, *J. Am. Chem. Soc.*, 98, 7132, 1976.

159. **Liu, M. T. H., Palmer, G. E., and Chishti, N. H.**, Photolysis and thermolysis of 3-*n*-butyl-3-phenyldiazirine, *J. Chem. Soc. Perkin Trans. II*, 53, 1981.

160. **Liu, M. T. H. and Jennings, B. M.**, Consecutive reactions in the thermal decomposition of phenylalkyldiazirines, *Can. J. Chem.*, 55, 3596, 1977.

161. **Liu, M. T. H. and Chien, D. H. T.**, Effects of solvent on the 1,2-hydrogen shift to a divalent carbon, *Can. J. Chem.*, 52, 246, 1974.

162. **Bradley, G. F., Evans, W. B. L., and Stevens, I. D. R.**, Thermal and photochemical decomposition of cycloalkanespirodiazirines, *J. Chem. Soc. Perkin Trans. II*, 1214, 1977.

163. **Moss, R. A., Fantina, M. E., and Munjal, R. C.**, Intermolecular additions to cyclobutylchlorocarbene, *Tetrahedron Lett.*, 1277, 1979.

164. **Moss, R. A. and Wetter, W. P.**, Cyclopropylphenylcarbene — thermal control of intermolecular addition, *Tetrahedron Lett.*, 22, 997, 1981.

165. **Békhazi, M., Risbood, P. A., and Warkentin, J.**, Generation and chemical properties of dicyclopropylcarbene. Ring expansion, chlorine abstraction, C-H insertion, and alkene addition reactions, *J. Am. Chem. Soc.*, 105, 5675, 1983.

166. **Baron, W. J., DeCamp, M. R., Hendrick, M. E., Jones, M., Jr., Levin, R. J., and Sohn, M. B.**, Carbenes from diazo compounds, in *Carbenes*, Vol. 1, Jones, M., Jr. and Moss, R. A., Eds., Wiley-Interscience, New York, 1973, chap. 1.

167. **Kirmse, W. and Streu, J.**, 1-Methyl-7-norbornylidene. Selective alkyl shifts of carbenes, *Chem. Ber.*, 117, 3490, 1984.

168. **Warner, P. and Chu, I.-S.**, The Skattebøl rearrangement: evidence for the carbene to carbene mechanism, *J. Org. Chem.*, 49, 3666, 1984.

169. **Sheridan, R. S. and Kasselmayer, M. A.**, Infrared spectrum and photochemistry of methoxychlorocarbene, *J. Am. Chem. Soc.*, 106, 436, 1984.

170. **West, P. R., Chapman, O. L., and LeRoux, J.-R.**, Cyclohepta-1,2,4,6-tetraene, *J. Am. Chem. Soc.*, 104, 1779, 1982.

171. **Stierman, T. J. and Johnson, R. P.**, Cumulene photochemistry: photorearrangement of 1,2-cyclononadiene to a bicyclic cyclopropene, *J. Am. Chem. Soc.*, 105, 2492, 1983.

172. **Wentrup, C. and Becker, J.**, Synthesis of 1-azaazulene and benz[*a*]azulene by carbene rearrangement, *J. Am. Chem. Soc.*, 106, 3705, 1984.

173. **Griller, D., Liu, M. T. H., and Scaiano, J. C.**, Hydrogen bonding in alcohols: its effect on the carbene insertion reaction, *J. Am. Chem. Soc.*, 104, 5549, 1982.

174. **Griller, D., Liu, M. T. H., Montgomery, C. R., Scaiano, J. C., and Wong, P. C.**, Absolute rate constants for the reactions of some arylchlorocarbenes with acetic acid, *J. Org. Chem.*, 48, 1359, 1983.

175. **Liu, M. T. H., Chishti, N. H., Tencer, M., Tomioka, H., and Izawa, Y.**, Reactive intermediates in the photolysis and thermoloysis of 3-chloro-3-benzyldiazirine, *Tetrahedron*, 40, 887, 1984.

176. **Liu, M. T. H. and Subramanian, R.**, Termolecular trapping of benzylchlorocarbene by methanol, *J. Chem. Soc. Chem. Commun.*, 1062, 1984.

177. **Liu, M. T. H. and Subramanian, R.**, Reaction of benzylchlorocarbene with hydrogen chloride, *J. Org. Chem.*, 50, 3218, 1985.

178. **Warner, P. M. and Chu, I.-S.**, Carbene insertion into methanol: a case of reversible ylide formation, *J. Am. Chem. Soc.*, 106, 5366, 1984.

179. **Oikawa, S. and Tsuda, M.**, Origin of the temperature-dependent product variation in carbene reactions with alcohol, *J. Am. Chem. Soc.*, 107, 1940, 1985.

180. **Liu, M. T. H. and Kokosi, J.**, Transformation of diazirine to 1,3-dioxolane and 1,3-dithiolane, *Heterocycles*, 3049, 1985.

181. **Liu, M. T. H.**, The thermolysis and photolysis of diazirines, *Chem. Soc. Rev.*, 11, 127, 1982.

182. **Kirmse, W., Loosin, K., and Sluma, H.-D.**, Carbenes and the O–H bond: cyclopentadienylidene and cycloheptatrienylidene, *J. Am. Chem. Soc.*, 103, 5935, 1981.

183. **Kirmse, W., Chiem, P. V., and Henning, P. G.**, Carbenes and the O–H bond: 3-cyclopentenylidene. A novel approach to bis(homo-cyclopropenyl) cations, *J. Am. Chem. Soc.*, 105, 1695, 1983.

184. **Crow, W. E. and McNab, H.,** Synthetic applications of intramolecular insertion in arylcarbenes. I. *o*-Alkylphenylcarbenes and 1-(*o*-alkylphenyl)ethylidenes, *Aust. J. Chem.* 32, 89, 1979.

185. **Crow, W. D. and McNab, H.,** Synthetic applications of intramolecular insertion in arylcarbenes. II. *o*-Alkoxy-, dialkylamino-, and alkylthiophenylcarbenes, *Aust. J. Chem.*, 32, 99, 1979.

186. **Crow, W. E. and McNab, H.,** Synthetic applications of intramolecular insertion in arylcarbenes. III. Generation and rearrangement of aryloxycarbenes by intramolecular abstraction of hydrogen, *Aust. J. Chem.*, 32, 111, 1979.

187. **Crow, W. D. and McNab, H.,** Synthetic applications of intramolecular insertion in arylcarbenes. IV. *o*-alkoxy-, dialkylamino-, and alkylthio-1-phenylethylidenes, *Aust. J. Chem.*, 32, 123, 1979.

188. **Crow, W. D. and McNab, H.,** Synthetic applications of intramolecular insertion in arylcarbenes. V. *o*-benzyl-, phenylamino-, phenoxy-, and phenylthiophenylcarbenes, *Aust. J. Chem.*, 34, 1037, 1981.

189. **Tomioka, H. and Izawa, Y.,** Photolysis of aryldiazomethanes in alcoholic matrices. Temperature and host dependencies of phenylcarbene processes in alcohols, *J. Am. Chem. Soc.*, 99, 6128, 1977.

190. **Senthilnathan, U. P. and Platz, M. S.,** Determination of the absolute rates of decay of arylcarbenes in various low temperature matrices by electron spin resonance spectroscopy, *J. Am. Chem. Soc.*, 102, 7637, 1980.

191. **Giese, B., Lee, W.-B, and Neumann, C.,** Evidence for intermediates in the cycloaddition of singlet carbenes, *Agnew. Chem. Int. Ed. Engl.*, 21, 310, 1982.

Chapter 9

DIAZIRINES AS PHOTOACTIVATABLE REAGENTS IN BIOCHEMISTRY

Hagan Bayley

TABLE OF CONTENTS

I. INTRODUCTION

Several lengthy reviews are available on the applications of photoactivatable reagents in biochemistry and molecular biology (see Reference 1 and references therein). This discussion, which explains how diazirines have been used in these fields, is a companion to the one on aryl azides which appeared recently.[2] Both have been written with the hope that organic chemists and photochemists might be sufficiently stimulated to take up investigations of the several problems encountered by the biochemists who use photochemical reagents but have neither the time nor the expertise to solve them.

II. APPLICATIONS OF DIAZIRINES AS PHOTOACTIVATABLE REAGENTS IN BIOCHEMISTRY

Diazirines have been used in two major areas: first, as photoaffinity reagents to label receptors and, second, as reagents for investigating the organization of biological membranes. Photochemical reagents such as diazirines may be activated at will, and the intermediates formed from them are extremely reactive. The importance of these two characteristics will be clarified later. By contrast, the reactions of the conventional chemical reagents used in biochemistry are more sluggish and cannot be controlled.[3-5]

A. Photoaffinity Reagents

For the purposes of this discussion, all biological molecules which bind ligands are regarded as receptors. For example, proteins on cell surfaces that bind hormones are commonly regarded as receptors, but other proteins, such as enzymes and immunoglobulins, which bind substrates, effectors, and antigens, may be conveniently placed in the same category.

In a photoaffinity labeling experiment,[6,7] a close analog of a ligand is made. It contains a photoactivatable group and it is usually radiolabeled. This analog is allowed to bind to its receptor and it is then irradiated, yielding a reactive intermediate (a carbene in the case of a diazirine) that reacts covalently with the receptor.

Photoaffinity reagents may be used to:

1.   Identify a receptor in a complex biological mixture. After photoaffinity labeling, the proteins in the mixture are separated, for example by electrophoresis, and the radio-labeled polypeptide is identified.
2.   Identify the ligand binding site within a receptor molecule. After photoaffinity labeling, the purified receptor polypeptide is cleaved into fragments using enzymes or chemical reagents, and the radiolabeled fragments are identified.
3.   Permanently switch a receptor from one state to another.

B. Hydrophobic Reagents for Membranes

Photoactivatable reagents have been devised that partition into the hydrocarbon phase of biological membranes.[8,9] On activation these molecules react with those parts of membrane proteins that lie in the lipid bilayer. If radiolabeled reagents are used, analysis of the membrane proteins reveals which are present only on the surface of the membrane (peripheral proteins) and which are at least in part buried in the membrane (integral proteins). The portions of the integral proteins that are in the lipid bilayer can be determined by further analysis.

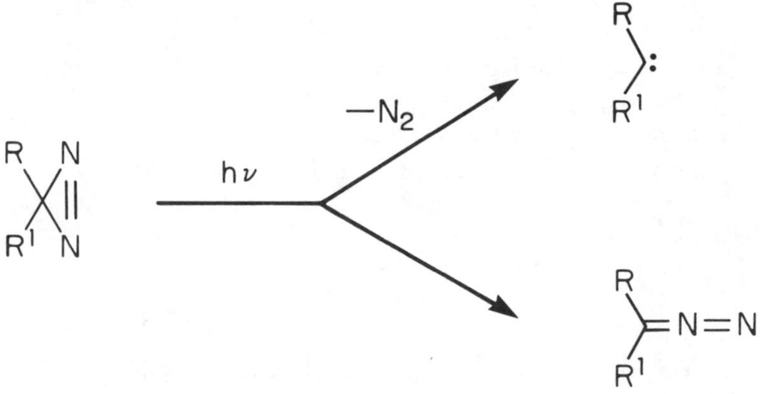

FIGURE 1.   Photochemistry of diazirines.

III. PROPERTIES REQUIRED OF PHOTOACTIVATABLE REAGENTS

Further details of the procedures outlined above are given later. The properties which superior photoactivatable reagents would be expected to have are as follows:

1.   They should be stable in the dark under physiological conditions.
2.   They should be efficiently activated by light outside the range of wavelengths which damages many biological preparations, viz., above 300 nm.
3.   They should form highly reactive intermediates on irradiation (the biological target might be quite unreactive or in a short-lived conformation).
4.   These intermediates should form *stable* covalent adducts with the biological targets, which are usually proteins.
5.   Finally, it would be advantageous if the reagent were easy to synthesize.

Molecules that have been used as photoaffinity reagents or for investigating membrane structure include diazo compounds, aryl azides, α,β-unsaturated ketones, aryl halides, nitroaryl compounds, unmodified nucleotides, psoralens, *N*-nitroso compounds, and diazonium salts.[1] With regard to the criteria listed above, diazirines compare favorably in almost all respects, as the following discussion of the relevant photochemistry and chemistry demonstrates. Throughout this chapter I draw comparisons with aryl azides in particular. Because they are especially easy to prepare, they are the most commonly used photochemical reagents.[2]

IV. PHOTOCHEMISTRY OF DIAZIRINES

All diazirines absorb at 350 to 380 nm (ϵ approximately 250) well away from the absorption maxima of proteins (280 nm) and nucleic acids (260 nm). In this regard they are excellent reagents. By comparison, unmodified aryl azides absorb maximally at 250 nm with little absorbance at 300 nm, although substitution with electron-withdrawing groups may be used to generate absorption bands in the visible.[7]

When diazirines are irradiated, two reactions predominate: loss of N_2 to yield a carbene and rearrangement to the linear diazo isomer (Figure 1).[10-12] The latter undesirable reaction occurs to the extent of 30 to 70%. Although diazo compounds can themselves be photolysed to yield carbenes the efficiency is relatively low above 300 nm. Furthermore, most diazo compounds can react with proteins in the dark. Therefore, the rearrangement converts a photoaffinity reagent to a conventional affinity reagent with all the associated drawbacks.[1-5] To circumvent this difficulty, two approaches can be used. First, diazirines yielding

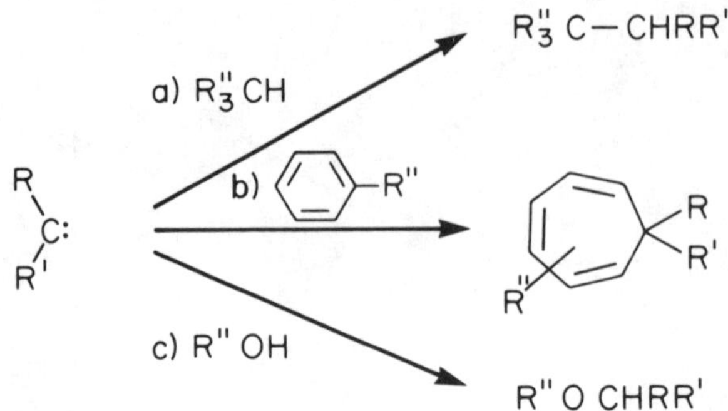

FIGURE 2.   Some important reactions of carbenes: (a) insertion into saturated hydrocarbons; (b) addition to unsaturated hydrocarbons, e.g., aromatics (this reaction proceeds via a cyclopropane intermediate); (c) reaction with nucleophiles, e.g., alcohols.

relatively inert diazo isomers have been employed.[12] Second, reagents that scavenge the diazo compound (such as thiols) can be added to the reaction medium.

The carbenes formed from diazirines are exceedingly reactive intermediates, as witnessed by other chapters in these volumes. They have the potential to react with all the functional groups found in biological molecules, including, importantly, those forming the amino acid side chains that make up the binding sites of receptors (Figure 2). Superficially, at least, the reactions of carbenes are simpler than those of nitrenes. In most cases, the products can be predicted (Table 1). In particular, complications arising from rearrangements of aryl carbenes to other reactive species seem to be unimportant in solution at room temperature. By comparison, under similar conditions, benzazirines and azacycloheptatetraenes feature strongly in the chemistry of aryl nitrenes (Figure 3).[13,14] The ability to predict the structures of many of the reaction products in turn allows us to appraise their stability. In the case of aryl azides, where the reaction products are unknown or possess unfamiliar chemistry, this is not generally the case.[2] It is important to know about the stabilities of the linkages formed between the reagent and the receptor, because the modified receptor is often subjected to harsh treatments during the analytical procedures that follow the labeling experiment (Table 1). For example, the esters formed when carbenes react with carboxyl side chains (found in Asp and Glu) might be cleaved in many of the acidic aqueous solvents used for chromatography of peptides and polypeptides, or by hydroxylamine which is used to cleave polypeptides at Asn-Gly bonds.

The classes of diazirines chosen for biochemical experiments and some comments on their properties are given in Table 2. Certain diazirines have been avoided because the carbenes formed from them rearrange to form inert molecules, e.g., dialkyl carbenes in general form alkenes. Others have been avoided because the reaction products would be unstable, e.g., α-haloaryl diazirines are relatively easy to make but many of the adducts formed between them and amino acid side chains would likely be easily hydrolyzed (Figure 4).

V. CHEMISTRY OF DIAZIRINES

A. Synthesis

Three strategies may be employed for the synthesis of photoaffinity reagents containing the diazirinyl group:

1.   The conversion of an existing functional group to a diazirine ring.
2.   The *de novo* synthesis of a ligand analog.
3.   The attachment of a diazirinyl group to a preexisting ligand using a bifunctional reagent.

Table 1

REACTIONS OF FUNCTIONAL GROUPS IN AMINO ACIDS WITH THE CARBENE R$_2$C

Group	Amino acid	Product	Stability of product[a]
CH	Ala, Val, Leu, Ile, Pro, Gly, etc.	C–CHR$_2$	Stable
Aromatic	Phe, Tyr, Trp, His	See Figure 2	Rearrangements in acid, for example, but reagent-receptor linkage remains intact
R′CH=CHR″	Trp	Cyclopropanes	Ring opened by strong acids and oxidants but linkage probably intact
CH$_3$S–	Met	$\overset{\oplus}{-}$S–CH$_3$ | CHR$_2$	Cleaved by thiols, about half the linkage should remain intact. Cleaved by elimination, e.g., by base
HOCHR′–	Ser, Thr, Tyr	R$_2$CHOCHR′–	Relatively stable
HSCH$_2$	Cys	R$_2$CH–S–CH$_2$–	Relatively stable
–S–S–	Cys-Cys	$\overset{\oplus}{-}$S–S | CHR$_2$	Facile –S$_2$– cleavage, reagent remains linked to the S atom
–COOH	Glu, Asp	–COOCHR$_2$	Unstable toward aqueous acid and base, and toward NH$_2$OH
–CONH$_2$	Gln, Asn	–C(OCHR$_2$)=NH	Very unstable toward hydrolysis
–NHR′	Lys, His, Trp	–NR′–CHR$_2$	Relatively stable
–NH–C(=NH)–NH$_2$	Arg	–NH–C(=NH)–NCHR$_2$	Relatively stable

[a]   Under conditions commonly used in protein chemistry.

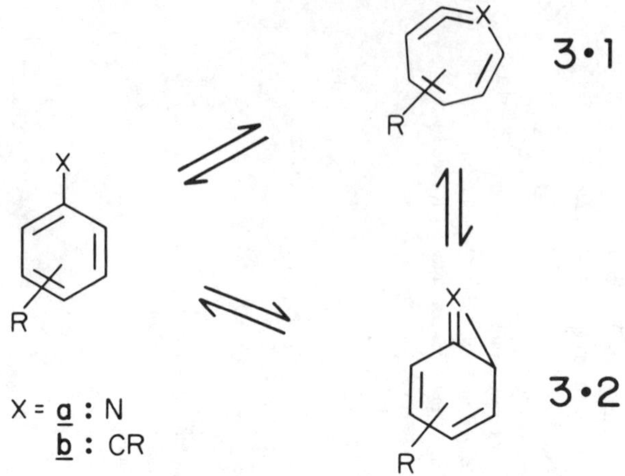

FIGURE 3.   Rearrangements of aryl carbenes and nitrenes to less reactive intermediates: **3.1a**, an azacycloheptatetraene; **3.2a**, a benzazirine; **3.1b**, a cycloheptatetraene; **3.2b**, a benzcyclopropene.

Table 2

	λ (ε)	Comments	Ref.
	350—400 nm (200—300)	Difficult synthesis; diazo rearrangement product reactive	10, 15, 17, 18, 24, 41, 45, 51, 64, 72, 86, 87
	353 nm (266)	Straightforward synthesis; diazo rearrangement product quite inert	12, 19, 20, 23, 26, 27, 31, 55, 65, 70, 71, 88
	372 nm (245)	Straightforward synthesis (no derivatives available yet); diazo rearrangement product very reactive	11, 21, 50, 56, 63, 66, 68

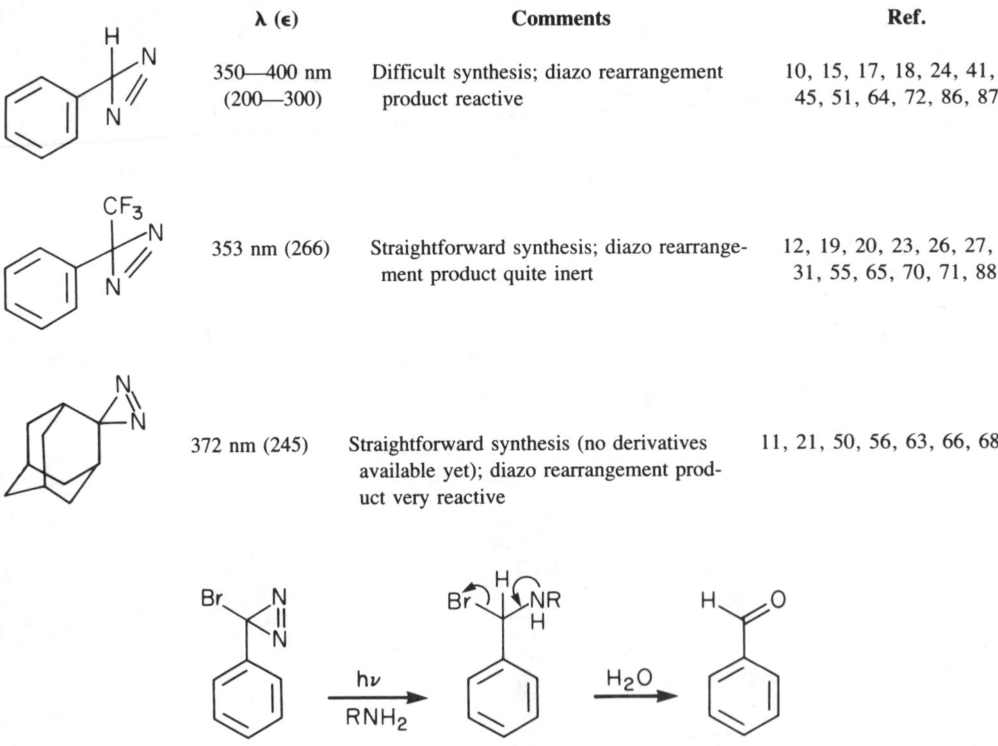

FIGURE 4.   The products formed when haloarylcarbenes react with nucleophiles may be unstable. Here a product that might be formed on reaction with a primary amine (e.g., the ε-amino group of lysine) is shown.

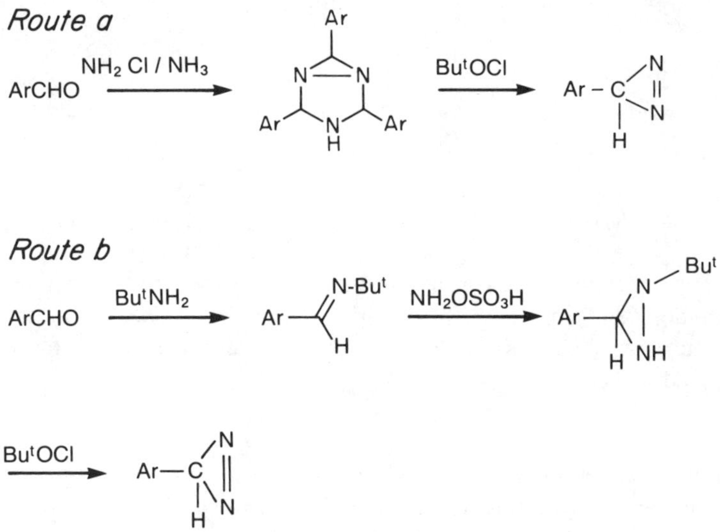

Route a

Route b

FIGURE 5.   Two routes to 3H-aryldiazirines from benzaldehydes.

In all cases, the problem of introducing a radioactive isotope, usually of high specific radioactivity, must be kept in mind.

Techniques for synthesizing diazirines are discussed fully in Chapter 3. Here, the methods that have proved practicable for making biological ligands are briefly reviewed.

1. 3H-3-Aryldiazirines

In their original investigations, Smith and Knowles[10,15] found two methods giving poor but usable yields of 3H-3-aryldiazirines. In route *a* (Figure 5) an aryl aldehyde is converted to a triazabicyclohexane by treatment with chloramine and ammonia. While this intermediate may be isolated when the aromatic ring is unsubstituted,[15,16] Such molecules are generally unstable. If treated at once with *t*-BuOCl, they are oxidized to yield diazirines in a 10 to 25% yield.[17,18] When electron-donating substituents are present, chlorination of the aromatic nucleus must be avoided by conducting the oxidation in the presence of pyridine.[18] As an alternative to *t*-BuOCl oxidation, an intermediate present during bicyclohexane formation may be trapped and converted to the diazirine with mercuric oxide.[15] Yields are only 1 to 4%.

Route *b* (Figure 5) may be used to produce aryldiazirines with electron-withdrawing substituents.[15] It is a simple procedure, as all the reactions are performed in a single flask, but yields are again low (3 to 15%).

2. 3-Trifluoromethyl-3-Aryldiazirines

3-Trifluoromethyl-3-aryldiazirines are more easily obtained than their 3H counterparts.[12] Tosyl hydrazones derived from trifluoromethylaryl ketones are converted to diaziridines with NH_3. The diaziridines are oxidized to diazirines using silver oxide. The overall yield from the ketone is 50 to 60% (Figure 6). Trifluoromethyl ketones may be prepared from aryl magnesium halides[12] or aryl lithiums.[19] The use of mesyl hydrazones instead of tosyl hydrazones may be advantageous.[20]

3. Adamantane Diazirine

Adamantane diazirine was synthesized as a hydrophobic reagent for membranes[11] by a method based on a general procedure for making dialkyl diazirines.[21] Adamantanone was

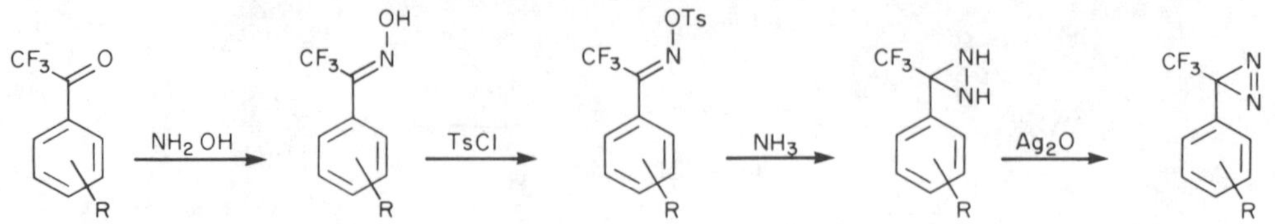

FIGURE 6.   Synthesis of 3-trifluoromethyl-3-aryldiazirines from trifluoroacetophenones.

converted to the corresponding diaziridine in 81% yield by treatment with ammonia followed by hydroxylamine-*O*-sulfonic acid. The diazirine was formed in near quantitative yield after oxidation of the diaziridine with acidic CrO_3. Functionalized adamantane diazirines have not been prepared as yet.

4. Examples

The conditions used in these syntheses are harsh, and consequently they have not been used often to convert existing functional groups in ligands to diazirines (method 1). In one example, the 7-keto group in ring B of a steroidal bile salt was converted to a diazirine ring.[22] The derivative gave easily detected labeling of bile salt binding proteins in hepatocytes,[23] although the extent of labeling was probably reduced by rearrangement of the photogenerated carbene to alkenes.

Biological specificity is not absolute and it is often possible to make ligand analogs that bind tightly to receptors (method 2). The synthesis of a diazirinyl retinal (Figure 7) illustrates the strategy that is employed.[24] Structure-activity studies with bacteriorhodopsin (see Section VI.A) revealed that the β-ionone ring of retinal (**7.1**; Figure 7) could be replaced by an aromatic ring (**7.2**; Reference 25, Figure 7). The diazirinyl aldehyde **7.3** was coupled by a Wittig reaction to an ylid containing the polyene chain of retinal. After conversion to the unlabeled analog, tritium was introduced by reduction of the aldehyde group with NaB^3H_4 followed by reoxidation.

A second example is the synthesis of an amino acid analog (Figure 8).[26] The diazirinyl-benzyl iodide **8.2** was prepared from the ketone **8.1** using the procedure outlined in Figure 6. Alkylation at C2 of a protected glycine derivative yielded the protected amino acid **8.3**, which was converted to the aromatic amino acid analog **8.4**. A second equally convenient route to **8.4** has been described.[27] The new amino acid has been resolved,[26,27] radiolabeled with tritium at the 2-position,[26] and converted to a biologically active enkephalin analog (using *t*-Boc protection and mixed anhydride coupling).[27] In the latter case at least, the diazirinyl group was not bulky enough to inhibit binding to enkephalin receptors (Phe was replaced by **8.4** in the pentapeptide Tyr-Gly-Gly-Phe-Leu).

The third method of making a photoaffinity reagent is to append a photoactivatable group to an existing ligand using a bifunctional reagent (Table 3). In our case, the first functional group would be a diazirine and the second a group capable of forming a linkage with the ligand. Bifunctional reagents can be used to modify small organic ligands, but they are most useful for modifying peptides or proteins (which can themselves be ligands for receptor proteins). For example, derivatives of the cyclic peptides phalloidin and antamanide have been produced by coupling a diazirinylbenzoic acid (Table 3, c) to them either via the N-hydroxysuccinimide ester, or by the mixed anhydride method.[28,29]

Compared with the profusion of bifunctional reagents containing arylazido groups,[1,2,30] few containing the diazirinyl group have been made. When such reagents are to be attached to proteins (i.e., polypeptides in excess of 5000 daltons), attention must be given to several important features:[1,2]

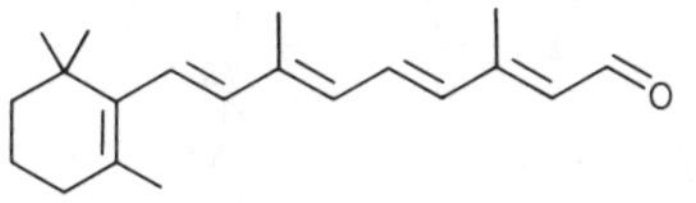

FIGURE 7. Synthesis of a diazirinyl analog of retinal **7.1**. (i) The diazirinylaldehyde **7.3** (the diazirine ring was constructed on a protected intermediate by the procedure shown in Figure 5 (route *a*)) was converted to the retinoic acid ester **7.4** by a Wittig reaction. (ii) The ester was converted to the radiolabeled diazirinyl retinal **7.5** in 4 steps: (a) Bu^i_2AlH; (b) MnO_2, to yield the unlabeled analog; followed by (c) NaB^3H_4; (d) MnO_2. The analog **7.2** functions as a cofactor for bacteriorhodopsin.

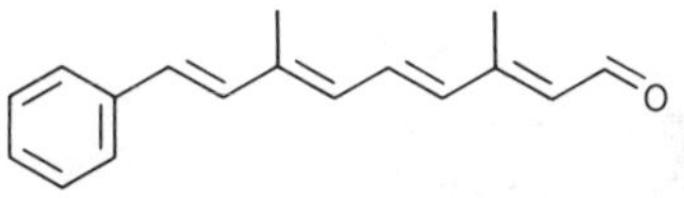

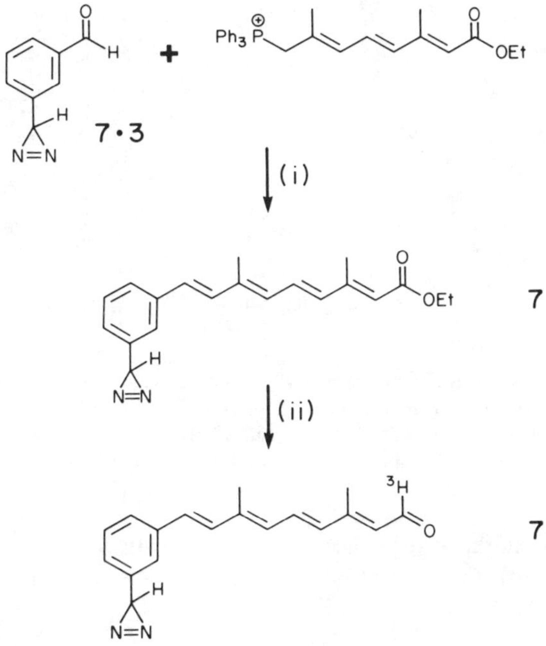

1.   The nature of the linkage between the reagent and the protein. It may be advantageous if the reagent only reacts with a specific functional group on the protein. The linkage should be stable during the photoaffinity labeling experiment, but it may be useful if it can be cleaved afterwards (see point 4).

2.   The solubility of the reagent. Most proteins are derivatized in aqueous solution.

3.   The length of the bifunctional reagent, i.e., the length of the spacer arm between the protein and the photoactivatable group, which must be able to reach its target but avoid more distant irrelevant molecules.

4.   The introduction of radioactivity. When a macromolecular photoaffinity reagent is used, the reagent-receptor adduct will have a greatly altered molecular weight. It might facilitate interpretation of the results to place a cleavable linkage in the spacer arm, and after photolysis transfer the radiolabel to the target receptor by cleavage of that linkage.

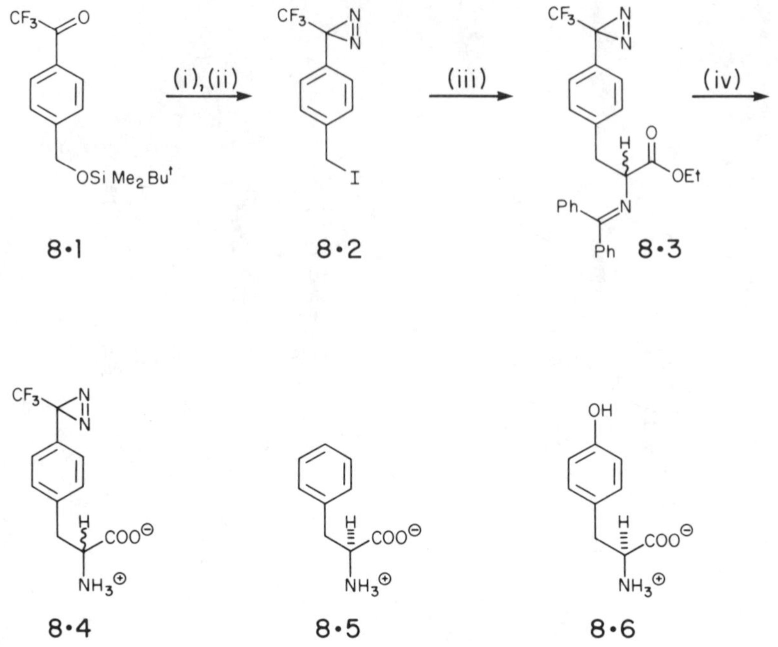

FIGURE 8.   Synthesis of a photoactivatable amino acid analog. (i) Construction of diazirine ring (see Figure 6); (ii) (a) HCl/MeOH, (b) $(PhO)_3MePI$; (iii) Ph_2:NCH_2COOEt (sodium salt of anion), EtOH-THF; (iv) (a) $CH_3COOH/H_2O/$ THF, (b) aqueous NaOH. **8.4** is an analog of phenylalanine, **8.5**, and of tyrosine, **8.6**.

Bifunctional molecules might also be used as cross-linking reagents for macromolecular systems, but so far, diazirines have not been adapted for this purpose.

B. Radiolabeled Reagents

When devising new syntheses of photoactivatable reagents, the organic chemist should realize that it will usually be necessary to introduce a radiolabel. Important aspects of this task are the specific radioactivity of the reagent, the position in the molecule at which the isotope is introduced, and the ease with which this can be performed. Where an existing functional group on a ligand is converted to a diazirine or a ligand such as a protein is modified with a bifunctional reagent, it is convenient to start with the radiolabeled ligand if it is readily available. A potential difficulty here is the separation of the diazirinyl group and the radioisotope in the molecule. Unless care is taken in the design of the reagent the isotope may be cleaved from the reagent-receptor adduct during work-up after the labeling experiment (see the remarks on the stability of the reagent-receptor bond). With this in mind, the isotope is usually placed close to the photoactivatable group when *de novo* syntheses of photoaffinity reagents and bifunctional reagents are carried out. At the least no easily cleaved bonds are placed between the two.

High specific radioactivity is often required, because many interesting receptors are present at extremely low concentrations in biological preparations. For example, ^{14}C might be placed in the diazirine group itself. Besides the synthetic difficulties, the low specific activity of carrier-free ^{14}C (62.4 mCi/matom) would make it useless for many applications. Virtually all ligands of interest contain hydrogen. It is often practicable to replace this atom with the inexpensive and relatively short-lived isotope ^3H. This isotope was used in the synthesis of the retinal analog (Figure 7: use of NaB^3H_4)[24] to prepare the radiolabeled amino acid analog

Table 3

		Comments	Ref.
a	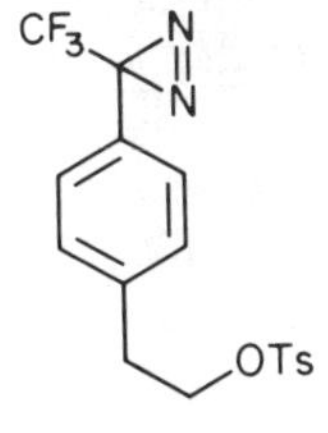	For attachment to nucleophilic groups but quite susceptible to elimination to form a styrene	12, 19, 87
b	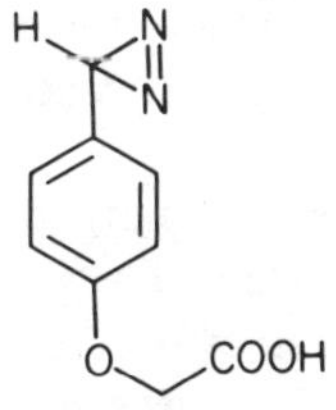	For attachment to $-NH_2$ groups either using a coupling agent or via an active ester; difficult synthesis	15
c	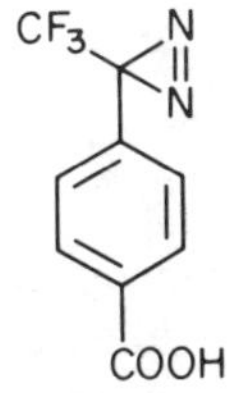	For attachment to NH_2 groups; see b; synthesis less demanding	19
d	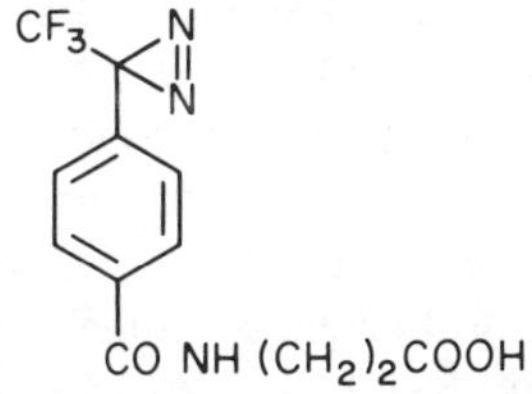	As in c but with a spacer arm	29
e	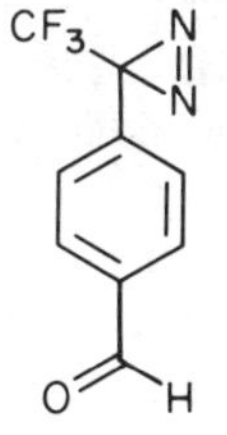	For attachment to $-NH_2$ by reductive alkylation using $[^3H]NaCNBH_3$	95

(Figure 8: use of $^3H_2O)^{26}$ and to prepare the phalloidin and antamanide analogs (use of $^3H_2$ gas).[28] The diazirine group, though stable toward most reagents, is susceptible to catalytic hydrogenation and in the latter case the isotope was introduced into the molecule before the diazirine ring.[28] ^{125}I, a second short-lived isotope, is also very useful. The hydrophobic reagent for membranes, trifluoromethyliodophenyl diazirine, is most conveniently prepared as the ^{125}I derivative.[23,31] Further details and lead references on isotopic syntheses may be found in Reference 1. A vital point is that the isotope should be introduced toward the end of a synthesis because expense and safety dictate that the operation be conducted on a small scale.

C. Stability of Diazirines

For many applications, the diazirinyl group must be unaffected by the harsh conditions used in organic synthesis. Further, the group should, of course, be stable under physiological conditions. Fortunately, this highly strained three-membered ring shows remarkable kinetic stability. In the synthesis shown in Figure 7, a 3H-diazirine is exposed to a phosphorus ylid, di-isobutyl aluminum hydride, MnO_2, and $NaBH_4$. In the synthesis of the amino acid shown in Figure 8, a trifluoromethylaryl diazirine is exposed to $(\phi O)_3MePI$, a carbanion, as well as strong acid and base. Acetic anhydride was used to acetylate the amino acid before resolution using the enzyme acylase. Cyclohexanespirodiazirine is stable toward transition metal salts, methyl iodide, peracetic acid, BF_3, Br_2, ϕ_3P, $LiNMe_2$, CH_2N_2, acidic dichromate, Ag_2O, acid chlorides, and alkaline hypobromite.[16] Other commonly used photoactivatable reagents, notably diazo compounds and aryl azides, do not exhibit such extraordinary chemical stability.[1] Therefore, a diazirine is usually the best photoactivatable group to accomplish a multistep organic synthesis.

Not surprisingly, diazirines are stable under physiological conditions. A particular drawback of diazo compounds and aryl azides is that they react rapidly with thiols,[32-35] commonly used to maintain reducing conditions in biological buffers. Aryl azides are particularly susceptible to dithiols such as dithiothreitol.[33] Trifluoromethylphenyldiazirine is stable in 100 mM dithiothreitol at pH 9.8.[12]

VI. PHOTOAFFINITY LABELING

A. Description of a Labeling Experiment

Photoaffinity labeling may be used to identify a receptor, to map the binding site of a ligand within a receptor, and to switch a receptor into a particular state. When a receptor is identified by photoaffinity labeling, the radiolabeled reagent is first allowed to bind to receptors in a complex biological preparation, e.g., whole cells or tissues, or a subcellular fraction. The mixture is then photolysed. Provided that nonspecific labeling is minimal, the extent of labeling is high, and the covalent bond formed between the receptor and the reagent is stable, the receptor usually can be identified. By "identified" I mean that approximations of the molecular weight of the receptor and its isoelectric point could be obtained by electrophoretic techniques.* These values would be useful if attempts were made to purify the receptor. Awareness of the molecular weight would aid in the choice of gel filtration media, and the isoelectric point would be helpful if fractionation techniques based on charge were to be used, e.g., ion-exchange chromatography or chromatofocusing. Photoaffinity labeling can be used to follow the purification of a protein. The protein can either be labeled before separation or individual fractions can be labeled and analyzed.[37,38]

Receptors can be switched from one state to another by photoaffinity reagents. For example, enzyme active sites have been blocked by covalent attachment of substrate analogs,[39] and cell-surface receptors have been induced to internalize with covalently attached hormone analogs.[40]

These aspects of photoaffinity labeling will not be discussed further here. To give a deeper insight into the problems involved a more detailed discussion of an experiment in which a binding site within a receptor was mapped[24] is presented below.

Bacteriorhodopsin is an integral membrane protein of 27,000 daltons (248 amino acid residues) from *Halobacterium halobium*. It acts as a light-driven proton pump. On absorption of a photon a proton (or perhaps two) is translocated across the cell membrane from the inside to the outside of the bacterium. The chromophore is retinal (**7.1**; Figure 7) which is covalently attached through the aldehyde group as a Schiff's base to Lys 216 of the protein.[41]

---

* For examples in which diazirines are used, see References 23 and 36. For other examples, see References 1 and 2.

It was of interest to determine where the opposite end of the retinal is located with respect to the protein. For that purpose the diazirinyl analog **7.5** was synthesized (Figure 7). Because the light sensitive analog could not be used in a light-dependent proton translocation assay a simpler aromatic analog **7.2** where the diazirinyl group was replaced by an H atom was incorporated into the apoprotein and shown to be active.[25] It was then assumed that the closely related diazirine would also be active, if an assay could be performed.* The natural chromophore was removed from the protein and replaced with the analog which formed a pigment with a stoichiometry of 1:1. It was shown that the analog, like retinal itself, was attached to Lys 216 by reducing the Schiff's base linkage under carefully defined conditions, isolating a peptide containing the attached analog and sequencing it by Edman degradation.

Having completed these experiments to demonstrate that the analog behaves similarly to retinal, photoaffinity labeling was carried out. The bacteriorhodopsin-analog complex was irradiated at 356 nm until no further radiolabel was incorporated into the protein. The Schiff's base linkage of a preparative sample was cleaved with hydroxylamine (to form an oxime) and 32% of the radiolabel remained covalently attached to the protein through linkages formed upon photolysis of the diazirine. The protein was then cleaved with chymotrypsin into two pieces, C2 (residues 1 to 71) and C1 (residues 72 to 248). These hydrophobic peptides were separated by gel filtration on Sephadex LH60 in organic solvent (88% HCO_2H/ EtOH(3:7)). Only C1 carried radiolabel. The peptide C1 was further fragmented by cleavage at methionine with cyanogen bromide in 70% HCO_2H. The CNBr fragments were separated by gel filtration, again on LH60, by gel electrophoresis or by thin layer chromatography. Radiolabel was found mainly in a peptide called CNBr-9 (residues 154 to 209). Peptide C1 was also cleaved at tryptophan with iodosobenzoic acid. The fragments were separated on LH60, followed by HPLC on a reversed-phase column eluted with an EtOH gradient in 5% HCO_2H. Most of the radiolabel was found in a peptide comprising residues 190 to 248. The overlap with CNBr-9 implied that the site of attachment of the photoaffinity reagent was between residues 190 and 209. This was confirmed by automated sequence analysis of the iodosobenzoic acid peptide. The major fraction of radioactivity eluted at cycles corresponding to Ser 193 and Glu 194. These data allowed several conclusions to be drawn concerning the structure and function of bacteriorhodopsin.[24] For example, they confirmed the idea[42] that the large red shift observed when retinal binds to the protein is due to a negative charge (Glu 194) in the vicinity of the β-ionone ring.

Clearly, the success of this experiment depended upon the stability of the reagent-receptor bond during the harsh manipulations required to discover the site of attachment. Some losses of radiolabel were noted at various stages, particularly after iodosobenzoic acid treatment, but this loss was shown to be due to partial oxidation of the oxime group carrying the tritium. Surprisingly, perhaps, the ester linkage to Glu 194 remained largely intact throughout the analysis.

B. Problems Arising During Photoaffinity Labeling

Several difficulties can arise in photoaffinity labeling experiments, including:

1. Nonspecific labeling
2. A low extent of labeling
3. Instability of the reagent-receptor bond
4. Instability of the reagent in the dark

---

* This aspect of the case is unusual. In general, a strong advantage of photoaffinity reagents over conventional affinity reagents is that they can be assayed for activity, in the dark, before photolysis.[1,2] Conventional affinity reagents begin reacting with the receptor immediately after mixing.[3-5]

1. Nonspecific Labeling
a. Causes

Nonspecific labeling occurs when a photoaffinity reagent reacts at locations other than its binding site within a receptor. Obviously, nonspecific labeling can lead to highly misleading results and hence great care must be taken to detect it, and where possible eliminate it. Two situations can lead to the problem: the binding site may have too weak an affinity for the reagent, or the residues in the binding site may be unreactive toward the intermediate generated from the reagent.

When a photoaffinity reagent binds only weakly to a receptor, it may be necessary to use it at high concentrations to bring the site close to saturation and thus produce detectable labeling. The consequent high concentration of free reagent can result in nonspecific labeling at weak binding sites or during random collisions. This situation can often be prevented by designing the reagent carefully, keeping structural alterations of the parent ligand to a minimum by mimicking its shape (Figure 7), or by making a minimal addition to its size (Figure 8). When photoactivatable groups are appended to preexisting ligands, their large size can often inhibit binding. Attempts have been made to reduce the bulk of the appended group but none have succeeded so far (see Section VIII). Of course, when a photoaffinity reagent is covalently bound to a receptor before irradiation (as in the case of bacteriorhodopsin) nonspecific labeling cannot occur.

A photoaffinity reagent may not react at the binding site under investigation if the intermediate generated from it is relatively unreactive. Under these circumstances the intermediate, even one which is quite tightly bound to a receptor, may live long enough to dissociate and react with another protein that happens to carry a particularly reactive functional group. For example, a ligand binding site might consist only of hydrocarbon residues. An aryl nitrene generated at such a site may dissociate (or rearrange to an even less reactive intermediate [Figure 3] and then dissociate), and react elsewhere with, say, a sulfhydryl group.[43] Some intermediates react so sluggishly that they may dissociate and reassociate several times before reacting. This process has been termed pseudophotoaffinity labeling.[44] Although the correct receptor may be labeled under such circumstances, all the problems associated with conventional affinity labeling may arise.[3-5]

Diazirines were introduced specifically to circumvent the problem of reactivity.[10,15] Indeed, in a direct comparison with aryl azides they have been shown to produce less nonspecific labeling.[45] An ideal photoaffinity reagent should be converted to a highly reactive intermediate on photolysis and not give rise to less reactive species either directly or by rearrangement of the intermediate. The carbenes generated from diazirines are certainly highly reactive and in general fulfill the first criterion.

Unfortunately, the diazirines used initially did produce relatively unreactive but far from inert byproducts,[10,11,15] viz., the linear diazo isomers (Figure 1 and Table 2). A useful advance was made when Brunner and colleagues[12] introduced trifluoromethylaryldiazirines. When these molecules are irradiated, the linear diazo isomers formed are quite inert. For example, they are stable toward 0.1 M CH_3COOH in cyclohexane at room temperature for 24 hr.[12] The reactivity of the linear isomers toward the various functional groups present in biological molecules (e.g., Table 1) has not been investigated systematically.

b. Detection and Prevention

Two important experiments are normally done when there is concern about nonspecific labeling. The first is a protection experiment. True photoaffinity labeling should be prevented by the presence of a strong ligand for the receptor (perhaps the natural ligand), while nonspecific labeling will remain unaffected. This is a useful way to detect nonspecific labeling. Scavengers reacting with unbound reactive intermediates can prevent nonspecific labeling. It may be essential to use scavengers when the receptor in question is present in

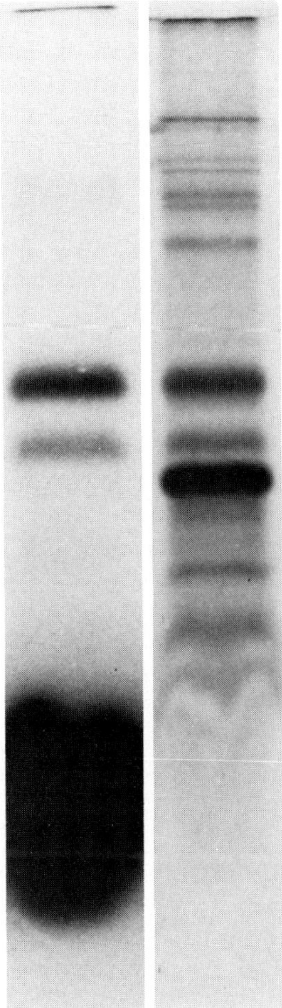

CHAPTER 9, FIGURE 10. Myelin membranes labeled with the hydrophobic reagent 3-(trifluoromethyl)-3-(m-[^{125}I]iodophenyl)diazirine. (a) Autoradiogram of a sodium dodecyl sulfate-polyacrylamide gel which separates the proteins in the membranes by molecular weight. The two bands are major and minor proteolipids, integral membrane proteins which have been labeled by the reagent. The dark area at the bottom of the gel is lipid that reacted with the reagent. (b) Coomassie blue stained gel showing all the proteins in the membranes. Beneath the two bands corresponding to the proteolipids is a third intensely stained band corresponding to the basic protein, the major peripheral protein which is unlabeled by the hydrophobic reagent. (P. Brophy & H. Bayley, unpublished.)

very low amounts, and nonspecific labeling is an appreciable fraction of the total. Several classes of molecules have been used as scavengers,[11,43,45,46] but thiols have proved most successful. Presumably, they can react with electrophilic intermediates by direct nucleophilic attack, and with radicals or triplet intermediates by hydrogen atom transfer (the SH bond is quite weak). All of these intermediates are found in carbene chemistry. Further consideration should be given to designing scavengers to match the properties of particular photoaffinity reagents. In the case of diazirines it would be useful to mop up the linear diazo compounds produced from them. Here, thiols again may be useful because they react with diazo compounds[35] but not with diazirines.[11,12]

Additional methods of dealing with nonspecific labeling include the removal of the unbound reagent before photolysis and washing the sample immediately after photolysis.[1,2] The design of the reagent may again influence the feasibility of using these methods. Often an unbound reagent cannot be removed, because exchange between free and bound ligand is rapid, but reagents that bind very tightly usually dissociate slowly. If the byproducts of photolysis are relatively unreactive (for example, diazoaryl compounds), it should be possible to carry out a rapid photolysis and then remove the unreacted molecules, e.g., by gel filtration.

2. Low Extent of Labeling

It is usually important to attain a high extent of labeling in a photoaffinity labeling experiment. Observed values range from very low indeed (less than 0.1%)[47] to almost 100% of the theoretical maximum.[48] When the goal is to identify a receptor in a mixture, a high extent of labeling will increase the limits of detection. When the goal is to identify the region of the receptor that binds the ligand, a high extent of labeling will aid in the analysis of labeled peptides derived from the polypeptide chain under examination. When the extent of labeling is low (say, approximately 1%), it can be extremely difficult, for lack of material, to identify (e.g., by amino acid analysis or by Edman degradation) labeled peptides that have been separated from their unmodified counterparts. When a receptor is "switched" from one state to another by photoaffinity labeling, an increase in the extent of labeling will increase the biological response.

Techniques such as repeated labeling can be used to improve the extent of labeling.[1] Much may be accomplished by correct design of the reagent. Again, the reagent should bind tightly, and the intermediate formed from it should be extremely reactive and not rearrange to less reactive species. In addition, it is advantageous if the reagent is activated efficiently (high ϕ and ϵ). Then labeling will take place before damage (e.g., photooxidation) to the biological sample. Finally, the bonds formed between the reagent and the receptor should be stable. An apparently low extent of labeling may be found where cleavage (e.g., hydrolysis) occurs soon after labeling, or while the sample is prepared for analysis.

3. Instability of the Reagent-Receptor Bond

Highly unstable reagent-receptor bonds are manifested in a low apparent extent of labeling. Moderately stable bonds (Table 1) may cleave during analysis when harsh reagents are used for peptide modification, cleavage, or degradation. Clearly, neither of these situations is satisfactory, but in the case of diazirines educated guesses may be made at the nature of the reaction products (Table 1). Diazirines that would yield highly unstable products such as α-haloaryl diazirines (Figure 4) should be avoided.

4. Instability of Reagents in the Dark

A final problem associated with photoaffinity reagents is their instability in the dark under physiological conditions. It has already been noted that diazirines are among the most stable reagents.

VII. STUDIES OF MEMBRANE PROTEIN TOPOGRAPHY

Photochemical reagents have proved extremely useful for examining the organization of proteins in biological membranes. Dividing membrane proteins into two classes, viz., integral and peripheral proteins, is a simple and useful concept.[49] Integral proteins are embedded in the lipid bilayer, indeed, they usually span it. Such proteins can be removed from the membrane only with difficulty; to do so it is usually necessary to dissolve the bilayer in detergent. Peripheral proteins are bound to the surface of the membrane, either to the lipid head groups or more often to integral proteins, and are more easily removed, for example, by treatment with buffers of high ionic strength.

A. Diazirines as Hydrophobic Reagents

Diazirines have been used extensively as hydrophobic reagents.[50-55] After equilibration with membranes, such reagents are located within the lipid bilayer. Upon photoactivation they react with the segments of integral membrane proteins found there. In principle they can therefore be used to answer three questions:

1. Which proteins are integral and which are peripheral?
2. Which segments of integral proteins are in the bilayer?
3. Which residues in these segments are directly exposed to the hydrocarbon core of the bilayer?

To ensure that a hydrophobic reagent reacts at the desired location, two factors must be considered. Again, the question of reactivity arises as it did with photoaffinity reagents. The majority of the amino acid side chains located within the bilayer are hydrophobic and hence inert. Therefore, only the most reactive intermediates will react with them. Carbene precursors are then the reagents of choice. Some comparisons with less reactive nitrenes will be drawn below.

The second consideration is that the reagent should be sufficiently hydrophobic to concentrate in the bilayer. Three classes of reagents have been used: simple hydrophobic reagents (i.e., organic molecules that dissolve in the bilayer,[11,31,56] amphipathic molecules,[57] and lipid analogs.[51,54] The simple hydrophobic reagents and the amphipathic molecules may be conveniently introduced into the membrane suspension in a small portion of organic solvent. A disadvantage is that they may diffuse in and out of the bilayer during the labeling experiment[11,46] (see Section VII.D). While the lipid analogs are tightly bound (if suitably designed, they will not exchange at all during the course of the experiment), they are more difficult to place in the bilayer and often the membranes must be dismantled (e.g., by dissolution in detergent) and reassembled with the analogs. In some cases it may be possible to introduce photoactivatable fatty acids into phospholipids in membranes biosynthetically.[58-60] Examples of simple hydrophobic reagents and phospholipid analogs are given in Figure 9.

B. Distinguishing Between Integral and Peripheral Proteins

There is clear evidence that carefully designed hydrophobic reagents are capable of distinguishing between integral and peripheral proteins.[31,56,61] Simple membranes containing relatively few proteins, such as those derived from human red blood cells or from myelin, have been photochemically labeled with radioactive hydrophobic diazirines such as [^{3}H]adamantane diazirine or trifluoromethyl-[^{125}I]iodophenyldiazirine (Figure 9). The proteins were separated by electrophoresis and then assayed individually for radioactivity.[31,56] In all cases, most of the radiolabel was found in the integral proteins (Figure 10; see glossy insert). As long as they are highly hydrophobic, less reactive intermediates from reagents

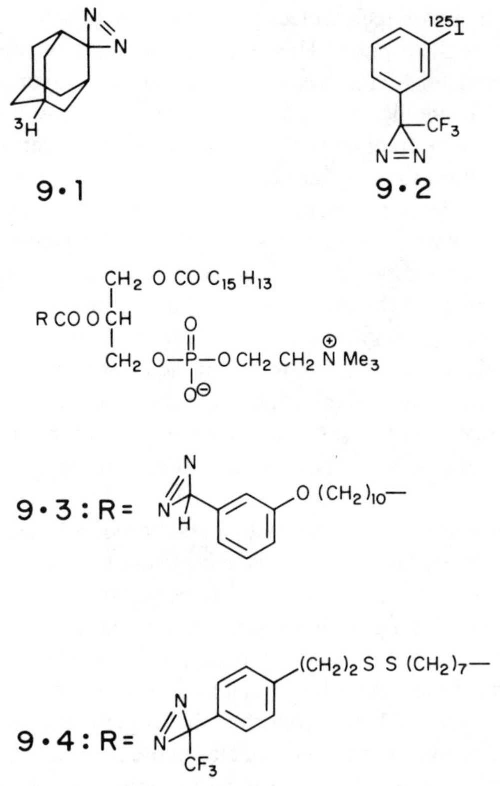

9·1   9·2

9·3 : R =

9·4 : R =

FIGURE 9.   Hydrophobic reagents for membranes. **9.1** and **9.2** are
"simple" hydrophobic reagents,[31,56] and **9.3** and **9.4** are lipid analogs.[51,87]

such as aryl azides are also capable of distinguishing between the two classes of proteins.[61] Weakly hydrophobic molecules that yield relatively unreactive intermediates, such as [^{3}H]phenyl azide, label both integral and peripheral proteins and are not useful as hydrophobic reagents.[56]

C. Labeling Those Segments of Integral Proteins that are in the Lipid Bilayer

Superficially, at least, integral membrane proteins fall into two groups: those that span the lipid bilayer once, and those that span it several times. It has proven fairly easy to show that proteins in the first group can be analyzed using hydrophobic reagents.[31,62-65] For example, glycophorin in red blood cells was labeled with [^{3}H]adamantane diazirine.[63] The protein was purified by affinity chromatography and dissected by chemical and proteolytic cleavage, which revealed that at least 75% of the tritium was located in the hydrophobic segment. This demonstrates remarkable selectivity, considering that this segment contains only 10% of the protein mass.

When more complex proteins are examined by this method, the protein chemistry becomes treacherous. In no instance have all the membrane-spanning segments of a complex protein been mapped along the polypeptide chain, although Nicholas has isolated five large hydrophobic peptides from the α-subunit of Na,K-ATPase after labeling them with [^{3}H]adamantane diazirine.[66] The analytical data on these peptides might be sufficient to identify them once the sequence of the polypeptide is known.

D. Reactivity and Selectivity

In most of the above, reactivity and selectivity have not been clearly distinguished. Of course, a highly reactive intermediate such as a carbene, capable of reacting with hydrocarbons, can still exhibit appreciable selectivity in its chemistry when given a *choice* of functional groups with which to react. This is in part a cause of nonspecific labeling with photoaffinity reagents, and forms the basis of the action of scavengers which can prevent nonspecific labeling. Hydrophobic reagents work because they are tightly bound to the lipid bilayer, i.e., they spend a large fraction of their time there. However, the simple hydrophobic reagents in particular partition rapidly between the hydrocarbon and aqueous phases.[67] If in the aqueous phase, they encounter groups that are, on average, approximately 1000 times more reactive than the available residues in the hydrocarbon phase, and they spend about 1% of their time outside the membrane, this is where they will react. For example, it was demonstrated at an early stage in the development of hydrophobic reagents that phenyl azide reacted with lipid bilayers to a very low extent when irradiated.[46] The carbenes generated from phenyl diazirine and adamantane diazirine, on the other hand, were capable of reacting with CH bonds in bilayers.[11] It is likely that the very reactive carbenes were short-lived compared to their residence time in the bilayer whereas the nitrene (or intermediates derived from it; Figure 3) was relatively long-lived, and therefore reacted with water or buffer components.

The difficulties that would arise if a substantial fraction of a hydrophobic reagent reacted from the aqueous phase can be largely eliminated by using highly hydrophobic reagents that are precursors to highly reactive intermediates. If necessary, aqueous scavengers such as glutathione (a tripeptide containing a −SH group) can be used.[46]

Within the membrane, however, selectivity may still present difficulties. For example, if one integral protein is ten times more highly labeled than a second it is impossible to say with certainty that the first protein has a proportionally larger part of its surface in the bilayer: it may simply react more readily with the particular photogenerated intermediate. Similarly, if a particular segment of an integral protein is more highly labeled than another, it cannot be said that the first segment presents a larger surface to the interior of the membrane. Inconsistencies that have come to light during studies of Na,K-ATPase may be manifestations of these difficulties.[66,68-70] The enzyme has two subunits, α and β. The distribution of label between the two varies depending upon the reagent used, as does the distribution of label along the polypeptide chain of the α-subunit.

It must be accepted that lightly labeled proteins could be either extremely reactive peripheral proteins or relatively inert integral proteins. An indication of which may be obtained by using a variety of hydrophobic reagents to see if the outcome is general, and by using scavengers to reduce the derivatization of peripheral proteins. The phospholipid reagents, although less convenient to use, should react exclusively within the bilayer. However, fatty acyl chains modified with bulky substituents can loop up to the membrane surface labeling reactive residues in the interphase region,[11,46,52,64] and these modified lipids can certainly bind to hydrophobic pockets in proteins. Reagents far more hydrophobic and indiscriminate will be needed to eliminate the problem of selectivity completely.

E. Distribution of Radiolabel within a Single Hydrophobic Segment

The problem of selectivity also arises when we ask which residues on a particular hydrophobic segment are exposed to the lipid bilayer and which are shielded by protein-protein (or segment-segment) interactions. Here, the problem has been solved, in part, by comparing the labeling of identical residues within a segment, e.g., by determining the distribution of radiolabel among the leucine residues in a peptide.[71,72] (The same method might be applied to the problems discussed above, but it would be very time consuming.)

For example, Hoppe and colleagues[71a] labeled the membrane-embedded section of ATP synthase from *Escherichia coli* with [^{125}I]trifluoromethyliodophenyl diazirine and subjected two of the subunits to extensive Edman degradation. After taking into account the relative reactivities of the amino acid side chains, they deduced that the N-terminal region of subunit

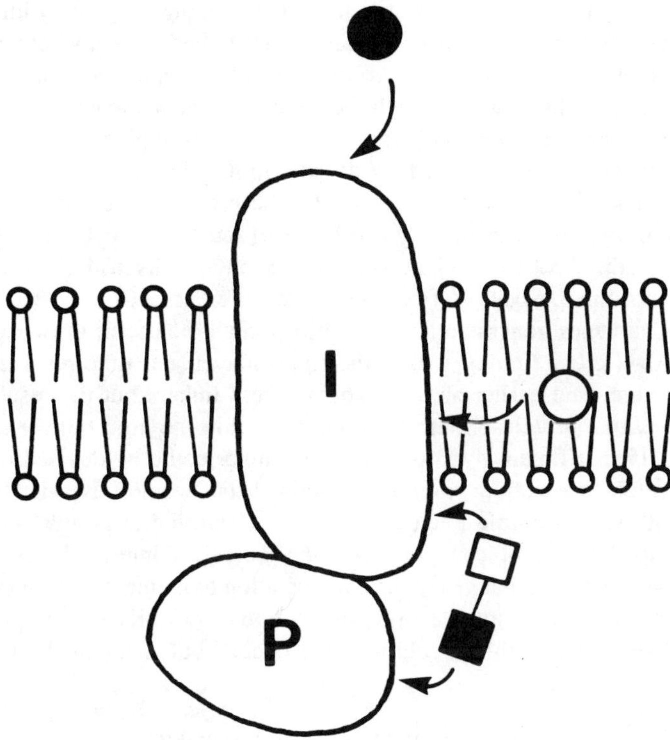

FIGURE 11. Section through a biological membrane, showing an integral protein (I), a peripheral protein (P), a hydrophilic surface-labeling reagent (●), a hydrophobic reagent (○), and a cross-linking reagent (□—■). (From Bayley, H. and Staros, J. V., *Azides and Nitrenes*, Scriven, E. F. V., Ed., Academic Press, New York, 1984, 433. With permission.)

b was buried in the bilayer with almost all of the side chains exposed to lipid, i.e., it had little interaction with nearby subunits or with itself. By contrast, a similar analysis of subunit *c* demonstrated that only two segments were exposed to the bilayer. The periodicity of the extent of labeling as a function of the cycle in the Edman degradation indicated that these segments were α-helices with only one face exposed to the lipids in the membrane. Aryl azide reagents could not be used for the analysis because they were far too selective:[73] only subunit *b* was derivatized and exclusively on a Cys residue.

F. Other Reagents for Membranes

Besides hydrophobic reagents, two other classes of reagents for investigating membrane protein topography have been devised (Figure 11). Water-soluble, surface labeling reagents provide information complementing that obtained with hydrophobic reagents. Because the surfaces of proteins exposed to the aqueous phase generally contain numerous reactive functional groups, there has been no pressing need to use such reactive species as carbenes to derivatize them. Conventional surface labeling reagents[74] and photochemical reagents based on aryl azide chemistry have served well.[75-78] Nevertheless, less discriminating reagents might be helpful in detecting segments of proteins exposed at the membrane surface that are not rich in reactive groups.[75]

Cross-linking reagents have been used to study protein-protein interactions at the surfaces of membranes.[79] For the most part, conventional bifunctional molecules with two electrophilic groups have been used. Based on the low extent of labeling found in most photoaffinity labeling experiments it might be expected that cross-linking reagents with two photoactivatable groups would work only poorly. Indeed, most studies of photochemical cross-linking reagents have centered on so-called heterobifunctional reagents,* in which one end of the molecule is an electrophilic group, and the other end is photoactivatable.[1,2,30] Carbene precursors have not been used in such experiments because high reactivity is not generally required. Cross-linking reagents are added directly to membranes and after the electrophilic group has reacted with exposed nucleophiles on the proteins, the photochemical group is activated. Protein-protein cross linking can be analyzed by 2D-gel electrophoresis, if the linkages can be readily broken. A cleavable bond, such as a disulfide, is often incorporated into the reagent for this purpose (for example, see Reference 80).

Nevertheless, carbenes generated from diazirines might be useful in two aspects of membrane biochemistry: cross linking within the lipid bilayer and time-resolved experiments. *Bis*-aryl azides were used earlier as hydrophobic cross linkers but the results were ambiguous.[81] It is likely that protein-protein interactions within the lipid bilayer are simpler than those outside it. If an efficient hydrophobic cross-linker could be devised new information about protein-protein interactions would be obtained that would be easier to interpret than that obtained with water-soluble reagents with two electrophilic groups, or with heterobifunctional reagents. Hydrophobic reagents yielding short-lived intermediates such as carbenes might be used to study rapid changes in the conformation of membrane proteins: for example, those that occur when water-soluble proteins assemble into bilayers. Experiments of this sort have been performed with a resolution of 15 sec,[57] but it should be possible to work on a far shorter time scale.

VIII. FUTURE DIRECTIONS

The deficiencies and caveats noted above suggest a need for several improvements to existing photochemical reagents and further that there are completely new directions that might be followed. Chemists and biochemists should participate in these developments, preferably as collaborators.

Controlling the reactivity of diazirines requires more attention. In many cases, highly reactive, indiscriminate intermediates must be generated and the existing reagents, while exceedingly reactive, are not optimally so. In other cases, more specificity could be desirable. For example, it would be much easier to analyze, at the level of protein chemistry, the sites labeled by a membrane surface reagent if it were selective. Of course, less information would be forthcoming as fewer groups on the membrane surface would be modified.

Further investigation of the photochemistry of diazirines should be performed, paying attention to the needs of the biochemist. Data on the quantum yields of photolysis of the diazirines suited to biochemical work should be gathered, and if necessary, means should be devised to increase the efficiency of photolysis (e.g., by energy transfer from absorbing groups on the receptor in the case of photoaffinity labeling[82]). The products formed when carbenes react with amino acid side chains should be investigated more carefully, especially for more complex residues (e.g., Trp). All investigations of photochemistry would best be conducted under relevant conditions, i.e., when possible at approximately pH 7 and in aqueous solution.

The stability of diazirines under physiological conditions, particularly their metabolic fate, has not been studied systematically and is worthy of further work. In some photoaffinity labeling experiments drug analogs have been administered to animals. After uptake and binding to receptors, isolated tissues were irradiated to activate the reagents.[83]

The size of photoactivatable groups is also a concern. Often, ligands no longer bind to

---

* Similar reagents are used in the synthesis of photoaffinity reagents (Table 3).

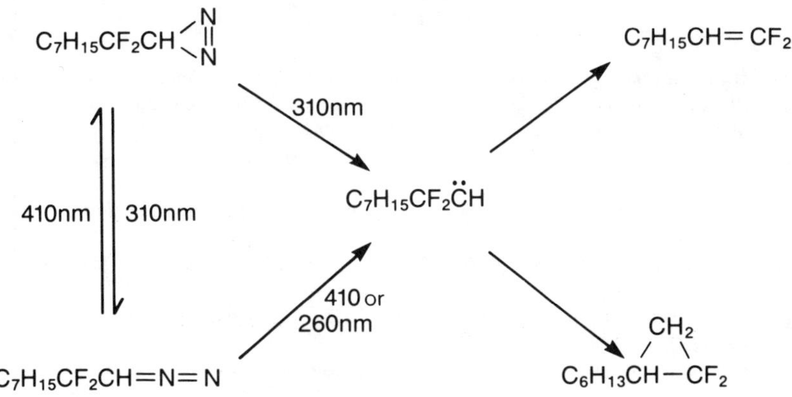

FIGURE 12.   Photochemistry of 3-(1,1-difluorooctyl)-3H-diazirine.

their receptors when bulky azidoaryl groups are appended to them using bifunctional reagents.[1] The same problem would no doubt arise if large bifunctional diazirines were used as alternatives. Erni and Khorana[84] attempted to reduce the size of the photoactivatable group by flanking the diazirine ring with $-CF_2-$ groups rather than with large aryl substituents. They were foiled, however, by intramolecular reactions (Figure 12).

Another area worthy of more research is in the analysis of the products of photoaffinity labeling and membrane labeling experiments. The protein chemistry involved is often tiresome and occasionally inconclusive. Usually, the analysis proceeds well up to the level of deciding which large peptides derived from a protein have been modified. Further analysis often yields a myriad of labeled peptides, many of which do not fractionate with their unmodified counterparts. A technique capable of dealing with such complex mixtures such as gas or liquid chromatography coupled to mass spectroscopy might be helpful. Hearne and Benisek[85] have used solid-phase photoaffinity reagents to recover active-site peptides. After covalent attachment of the receptor to the solid support by irradiation, the receptor is digested with protease until only a small peptide remains bound. The photoaffinity reagent is then cleaved from the support (for example, it might be attached through an ester linkage), releasing the active-site peptide (or peptides) which can be further purified and then identified by amino acid composition or sequence analysis.

Numerous opportunities remain for developing new classes of reagents. Bifunctional reagents, including cross-linking reagents, and membrane surface labeling reagents were mentioned earlier. Reagents that react at the interphase between the hydrocarbon domain and the buffer in membranes are being investigated,[86] and should be a useful addition to the classes of reagents used to investigate membrane protein topography (Figure 11). Attempts have been made to make reagents that react at a fixed depth in the bilayer[51,87] but problems have arisen because they are flexible and can react near the membrane surface if highly reactive amino acid side chains are available.[52,64,88] Rigid molecules might be used to greater effect for this purpose. A new and speculative way to produce polypeptide photoaffinity reagents would be by in vitro protein synthesis using photoactivatable amino acids.[26] Photoactivatable drugs have been used to treat various diseases, including cancer[89] and several sophisticated developments are being explored including drugs specific for certain cells.[90] Photoactivated irreversible inhibitors for key enzymes might also be investigated as therapeutic agents.

Time-resolved photochemical labeling, which requires that the photogenerated intermediates be very short-lived (or rapidly quenchable) is being investigated in several laboratories.[91-94]

REFERENCES

1. **Bayley, H.,** *Photogenerated Reagents in Biochemistry and Molecular Biology,* Elsevier, Amsterdam, 1983.
2. **Bayley, H. and Staros, J. V.,** Photoaffinity labeling and related techniques, in *Azides and Nitrenes,* Scriven, E. F. V., Ed., Academic Press, New York, 1984, 433.
3. **Lundblad, R. L. and Noyes, C. M.,** *Chemical Reagents for Protein Modification,* Vols. 1 and 2, CRC Press, Boca Raton, Fla., 1984.
4. **Plapp, B. V.,** Application of affinity labeling for studying structure and function of enzymes, *Methods Enzymol.,* 87, 469, 1982.
5. **Jacoby, W. B. and Wilchek, M., Eds.,** *Affinity Labeling, Methods in Enzymology,* Vol. 46, Academic Press, New York, 1977.
6. **Singh, A., Thornton, E. R., and Westheimer, F. H.,** The photolysis of diazoacetylchymotrypsin, *J. Biol. Chem.,* 237, PC3006, 1962.
7. **Knowles, J. R.,** Photogenerated reagents for biological receptor-site labeling, *Acc. Chem. Res.,* 5, 155, 1972.
8. **Klip, A. and Gitler, C.,** Photoactive covalent labeling of membrane components from within the lipid core, *Biochem. Biophys. Res. Commun.,* 60, 1155, 1974.
9. **Chakrabarti, P. and Khorana, H. G.,** A new approach to the study of phospholipid-protein interactions in biological membranes. Synthesis of fatty acids and phospholipids containing photosensitive groups, *Biochemistry,* 14, 5021, 1975.
10. **Smith, R. A. G. and Knowles, J. R.,** Aryldiazirines. Potential reagents for photolabeling of biological receptor sites, *J. Am. Chem. Soc.,* 95, 5072, 1973.
11. **Bayley, H. and Knowles, J. R.,** Photogenerated reagents for membrane labeling. II. Phenylcarbene and adamantylidene formed within the lipid bilayer, *Biochemistry,* 17, 2420, 1978.
12. **Brunner, J., Senn, H., and Richards, F. M.,** 3-Trifluoromethyl-3-phenyldiazirine: a new carbene generating group for photolabeling reagents, *J. Biol. Chem.,* 255, 3313, 1980.
13. **Scriven, E. F. V., Ed.,** *Azides and Nitrenes,* Academic Press, New York, 1984.
14. **Wentrup, C.,** *Reactive Molecules,* Wiley-Interscience, New York, 1984.
15. **Smith, R. A. G. and Knowles, J. R.,** Preparation and photolysis of 3-aryl-3H-diazirines, *J. Chem. Soc. Perkin Trans. II,* 686, 1975.
16. **Bradley, G. F., Evans, W. B. L., and Stevens, I. D. R.,** Thermal and photochemical decomposition of cycloalkanespirodiazirines, *J. Chem. Soc. Perkin Trans. II,* 1214, 1977.
17. **Radhakrishnan, R., Robson, R. J., Takagaki, Y., and Khorana, H. G.,** Synthesis of modified fatty acids and glycerophospholipid analogs, *Methods Enzymol.,* 72, 408, 1981.
18. **Leblanc, P. and Gerber, G. E.,** An improved synthesis of m-diazirinophenol, *Can. J. Chem.,* 62, 1767, 1984.
19. **Nassal, M.,** 4-(1-Azi-2,2,2-trifluoroethyl)benzoic acid, a highly photolabile carbene generating label readily fixable to biochemical agents, *Liebigs Ann. Chem.,* 1510, 1983.
20. **Brunner, J., Spiess, M., Aggeler, R., Huber, P., and Semenza, G.,** Hydrophobic labeling of a single leaflet of the human erythrocyte membrane, *Biochemistry,* 22, 3812, 1983.
21. **Schmitz, E. and Ohme, R.,** Cyclic diazo compounds. I. Preparation and reactions of diazirines, *Chem. Ber.,* 94, 2166, 1961.
22. **Kramer, W. and Kurz, G.,** Photolabile derivatives of bile salts. Synthesis and suitability for photoaffinity labeling, *J. Lipid Res.,* 24, 910, 1983.
23. **Wieland, T., Nassal, M., Kramer, W., Fricker, G., Bickel, U., and Kurz, G.,** Identity of hepatic membrane transport systems for bile salts, phalloidin, and antamanide by photoaffinity labeling, *Proc. Natl. Acad. Sci. U.S.A.,* 81, 5232, 1984.
24. **Huang, K. S., Radhakrishnan, R., Bayley, H., and Khorana, H. G.,** Orientation of retinal in bacteriorhodopsin as studied by cross-linking using a photosensitive analog of retinal, *J. Biol. Chem.,* 257, 13616, 1982.
25. **Bayley, H., Radhakrishnan, R., Huang, K. S., and Khorana, H. G.,** Light-driven proton translocation by bacteriorhodopsin reconstituted with the phenyl analog of retinal, *J. Biol. Chem.,* 256, 3797, 1981.
26. **Shih, L. B. and Bayley, H.,** A carbene-yielding amino acid for incorporation into peptide photoaffinity reagents, *Anal. Biochem.,* 144, 132, 1985.
27. **Nassal, M.,** 4'-(1-Azi-2,2,2-trifluoroethyl)phenylalanine, a photolabile carbene-generating analogue of phenylalanine, *J. Am. Chem. Soc.,* 106, 7540, 1984.
28. **Nassal, M., Buc, P., and Wieland, T.,** Synthese von [6-(L-2,6-Diamino-4-hexinsäure]Antamanid als vorstute eines ³H-markierten [6-Lysin]antamanids fur die photoaffinitäts-markierung, *Liebigs Ann. Chem.,* 1524, 1983.
29. **Wieland, T., Hollosi, M., and Nassal, M.,** *delta*-Aminophalloin, ein 7-analoges des phalloidins, und biochemisch nützliche, auch fluoreszierende derivate, *Liebigs Ann. Chem.,* 1533, 1983.

30. **Ji, T.,** Bifunctional reagents, *Methods Enzymol.,* 91, 580, 1983.
31. **Brunner, J. and Semenza, G.,** Selective labeling of the hydrophobic core of membranes with 3-(trifluoromethyl)-3-(m-[^{125}I]iodophenyl)diazirine, a carbene-generating reagent, *Biochemistry,* 20, 7174, 1981.
32. **Cartwright, I. L., Hutchinson, D. W., and Armstrong, V. W.,** The reaction between thiols and 8-azidoadenosine derivatives, *Nucleic Acids Res.,* 3, 2331, 1976.
33. **Staros, J. V., Bayley, H., Standring, D. N., and Knowles, J. R.,** Reduction of aryl azides by thiols: implications for the use of photoaffinity reagents, *Biochem. Biophys. Res. Commun.,* 80, 568, 1978.
34. **Bayley, H., Standring, D. N., and Knowles, J. R.,** Propane-1,3-dithiol: a selective reagent for the efficient reduction of alkyl and aryl azides to amines, *Tetrahedron Lett.,* 3633, 1978.
35. **Takagaki, Y., Gupta, C., and Khorana, H. G.,** Thiols and the diazo group in photoaffinity labels, *Biochem. Biophys. Res. Commun.,* 95, 589, 1980.
36. **Burgermeister, W., Nassal, M., Wieland, T., and Helmreich, E. J. M.,** A carbene-generating photoaffinity probe for *beta*-adrenergic receptors, *Biochim. Biophys. Acta,* 729, 219, 1983.
37. **Newman, M. J., Foster, D. L., Wilson, T. H., and Kaback, H. R.,** Purification and reconstitution of functional lactose carrier from *E. coli, J. Biol. Chem.,* 256, 11804, 1981.
38. **Shorr, R. G. L., Heald, S. L., Jeffs, P. W., Lavin, T. N., Strohsacker, M. W., Lefkowitz, R. J., and Caron, M. G.,** The *beta*-adrenergic receptor; rapid purification and covalent labeling by photoaffinity crosslinking, *Proc. Natl. Acad. Sci. U.S.A.,* 79, 2778, 1982.
39. **Staros, J. V. and Knowles, J. R.,** Photoaffinity inhibition of dipeptide transport in *E. coli, Biochemistry,* 17, 3321, 1978.
40. **Heidenreich, K. A., Brandenburg, D., Berhanu, P., and Olefsky, J. M.,** Metabolism of photoaffinity-labeled insulin receptors by adipocytes. Role of internalization, degradation, and recycling, *J. Biol. Chem.,* 259, 6511, 1984.
41. **Bayley, H., Huang, K. S., Radhakrishnan, R., Ross, A. H., Takagaki, Y., and Khorana, H. G.,** Site of attachment of retinal in bacteriorhodopsin, *Proc. Natl. Acad. Sci. U.S.A.,* 78, 2225, 1981.
42. **Nakanishi, K., Balogh-Nair, V., Arnaboldi, M., Tsujimoto, K., and Honig, B.,** An external point-charge model for bacteriorhodopsin to account for its purple color, *J. Am. Chem. Soc.,* 102, 7945, 1980.
43. **Nicolson, A. W., Hall, C. C., Strycharz, W. A., and Cooperman, B. S.,** Photoaffinity labeling of *E. coli* ribosomes by an aryl azide analogue of puromycin, *Biochemistry,* 21, 3797, 1982.
44. **Ruoho, A. E., Kiefer, H., Roeder, P. E., and Singer, S. J.,** Mechanism of photoaffinity labeling, *Proc. Natl. Acad. Sci. U.S.A.,* 70, 2567, 1973.
45. **Standring, D. N. and Knowles, J. R.,** Photoaffinity labeling of lactate dehydrogenase by the carbene derived from the 3-diazirino analog of nicotinamide adenine dinucleotide, *Biochemistry,* 19, 2811, 1980.
46. **Bayley, H. and Knowles, J. R.,** Photogenerated reagents for membrane labeling. I. Phenyl nitrene formed with the lipid bilayer, *Biochemistry,* 17, 2414, 1978.
47. **Niedel, J., Davis, J., and Cuatrecasas, P.,** Covalent affinity labeling of the formyl peptide chemotactic receptor, *J. Biol. Chem.,* 255, 7063, 1980.
48. **Walter, U., Uno, I., Liu, A. Y.-C., and Greengard, P.,** Identification, characterization, and quantitative measurement of cylic AMP receptor proteins in cytosol of various tissues using a photoaffinity ligand, *J. Biol. Chem.,* 252, 6494, 1977.
49. **Eisenberg, D.,** Three-dimensional structure of membrane and surface proteins, *Ann. Rev. Biochem.,* 53, 595, 1984.
50. **Bayley, H. and Knowles, J. R.,** Photogenerated hydrophobic reagents for intrinsic membrane proteins, *Ann. N.Y. Acad. Sci.,* 346, 45, 1980.
51. **Radhakrishnan, R., Gupta, C. M., Erni, B., Robson, R. J., Curatolo, W., Majumdar, A., Ross, A. H., Takagaki, Y., and Khorana, H. G.,** Phospholipids containing photoactivable groups in studies of biological membranes, *Ann. N.Y. Acad. Sci.,* 346, 165, 1980.
52. **Brunner, J.,** Labeling the hydrophobic core of membranes, *Trends Biochem. Sci.,* 6, 44, 1981.
53. **Bayley, H.,** Photoactivated hydrophobic reagents for integral membrane proteins, in *Membranes and Transport,* Vol. 1, Martonosi, A. N., Ed., Plenum Press, New York, 1982, 185.
54. **Robson, R. J., Radhakrishnan, R., Ross, A. H., Takagaki, Y., and Khorana, H. G.,** Photochemical crosslinking in studies of lipid-protein interactions, in *Lipid-Protein Interactions,* Vol. 2, Jost, P. C. and Griffiths, O. H., Eds., John Wiley & Sons, New York, 1982, 149.
55. **Brunner, J.,** Hydrophobic probes, *Methods Enzymol.,* in press.
56. **Bayley, H. and Knowles, J. R.,** Photogenerated reagents for membranes: selective labeling of intrinsic membrane proteins in the human erythrocyte membrane, *Biochemistry,* 19, 3883, 1980.
57. **Wisnieski, B. J. and Bramhall, J. S.,** Photolabeling of cholera toxin subunits during membrane penetration, *Nature (London),* 289, 319, 1981.
58. **Olsen, W. L., Schaechter, M., and Khorana, H. G.,** Incorporation of synthetic fatty acid analogs into phospholipids of *E. coli, J. Bacteriol.,* 137, 1443, 1979.

59. **Quay, S. C., Radhakrishnan, R., and Khorana, H. G.,** Incorporation of photosensitive fatty acids into phospholipids of *E. coli* and irradiation-dependent cross-linking of phospholipids to membrane proteins, *J. Biol. Chem.,* 256, 4444, 1979.

60. **Leblanc, P. and Gerber, G. E.,** Biosynthetic utilization of photoreactive fatty acids by rat liver microsomes, *Can. J. Biochem. Cell Biol.,* 62, 375, 1984.

61. **Bercovici, T., Gitler, C., and Bromberg, A.,** 5-[^{125}I]Iodonaphthyl azide, a reagent to determine the penetration of proteins into the lipid bilayer of biological membranes, *Biochemistry,* 17, 1484, 1978.

62. **Kahane, I. and Gitler, C.,** Red cell membrane glycophorin labeling from within the lipid bilayer, *Science,* 201, 351, 1978.

63. **Goldman, D. W., Pober, J. S., White, J., and Bayley, H.,** Selective labeling of the hydrophobic segments of intrinsic membrane proteins with a lipophilic photogenerated carbene, *Nature (London),* 280, 841, 1979.

64. **Ross, A. H., Radhakrishnan, R., Robson, R. J., and Khorana, H. G.,** The transmembrane domain of glycophorin A as studied by cross-linking using photoactivatable phospholipids, *J. Biol. Chem.,* 257, 4152, 1982.

65. **Spiess, M., Brunner, J., and Semenza, G.,** Hydrophobic labeling, isolation, and partial characterization of the amino-terminal membranous segment of sucrase-isomaltase complex, *J. Biol. Chem.,* 257, 2370, 1982.

66. **Nicholas, R. A.,** Purification of the membrane-spanning tryptic peptides of the α-polypeptide from sodium and potassium ion activated adenosine triphosphatase labeled with 1-tritiospiro(adamantane-4,3'-diazirine), *Biochemistry,* 23, 888, 1984.

67. **Turro, N. J., Zimmt, M. B., and Gould, I. R.,** Dynamics of micellized radical pairs, *J. Am. Chem. Soc.,* 105, 6347, 1984.

68. **Farley, R. A., Goldman, D. W., and Bayley, H.,** Identification of regions of the catalytic subunit of (Na,K)-ATPase embedded with the cell membrane. Photochemical labeling with [^{3}H]-adamantane diazirine, *J. Biol. Chem.,* 255, 860, 1980.

69. **Jorgensen, P. L., Karlish, S. J. D., and Gitler, C.,** Evidence for the organization of the transmembrane segments of (Na,K)-ATPase based on labeling lipid-embedded and surface domains of the α-subunit, *J. Biol. Chem.,* 257, 7435, 1982.

70. **Jorgensen, P. L. and Brunner, J.,** Labeling of intramembrane segments of the α-subunit and β-subunit of pure membrane-bound (Na,K)-ATPase with 3-trifluoromethyl-3-(m-[^{125}I]iodophenyl)diazirine, *Biochem. Biophys. Acta,* 735, 291, 1983.

71a. **Hoppe, J., Brunner, J., and Jorgensen, B. B.,** Structure of the membrane-embedded F_0 part of F_1F_0ATP synthase from *E. coli* as inferred from labeling with 3-(trifluoromethyl)-3-(m-[^{125}I]iodophenyl)diazirine, *Biochemistry,* 23, 5610, 1984.

71b. **Brunner, J., Franzusoff, A. J., Luscher, B., Zugliana, C., and Semenza, G.,** Membrane protein topology: amino acid residues in a putative transmembrane α-helix of bacteriorhodopsin labeled with the hydrophobic carbene-generating reagent 3-(trifluoromethyl)-3-(m-[^{125}I]iodophenyl)diazirine, *Biochemistry,* 24, 5422, 1985.

72a. **Takagaki, Y., Radhakrishnan, R., Gupta, C. M., and Khorana, H. G.,** The membrane-embedded segment of cytochrome b_5 as studied by cross-linking with photoactivatable phospholipids. I. The transferable form, *J. Biol. Chem.,* 258, 9128, 1983.

72b. **Takagaki, Y., Radhakrishnan, R., Wirtz, K. W. A., and Khorana, H. G.,** The membrane-embedded segment of cytochrome b_5 as studied by cross-linking with photoactivatable phospholipids. II. The non-transferable form, *J. Biol. Chem.,* 258, 9136, 1983.

73. **Hoppe, J., Montecucco, C., and Friedl, P.,** (^{125}I)Iodonaphthyl azide labeling selectivity a cysteine residue in the F_0 of the ATP synthase from *E. coli* is unsuitable for topographic studies of membrane proteins, *J. Biol. Chem.,* 258, 2882, 1983.

74. **Etemadi, A.-H.,** Membrane asymmetry. I. Methods for assessing the asymmetric orientation and distribution of proteins, *Biochim. Biophys. Acta,* 604, 347, 1980.

75. **Staros, J. V. and Richards, F. M.,** Photochemical labeling of the surface proteins of human erythrocytes, *Biochemistry,* 13, 2720, 1974.

76. **Staros, J. V., Haley, B. E., and Richards, F. M.,** Human erythrocytes and resealed ghosts. Comparison of membrane topology, *J. Biol. Chem.,* 249, 5004, 1974.

77. **Staros, J. V., Richards, F. M., and Haley, B. E.,** Photochemical labeling of the cytoplasmic surface of the membranes of intact human erythrocytes, *J. Biol. Chem.,* 250, 8174, 1975.

78. **Dockter, M. E.,** Fluorescent photochemical surface labeling of intact human erythrocytes, *J. Biol. Chem.,* 254, 2161, 1979.

79. **Peters, K. and Richards, F. M.,** Chemical cross-linking: reagents and problems in studies of membrane structure, *Ann. Rev. Biochem.,* 46, 523, 1977.

80. **Ji, T. H., Kiehm, K. D., and Middaugh, C. R.,** Presence of spectrin tetramer on the erythrocyte membrane, *J. Biol. Chem.,* 255, 2990, 1980.

81. **Mikkelsen, R. B. and Wallach, D. F. H.,** Photoactivated cross-linking of proteins within the erythrocyte membrane core, *J. Biol. Chem.,* 251, 7413, 1976.
82. **Goeldner, M. P., Hirth, G. G., Kieffer, B., and Ourisson, G.,** Photosuicide inhibition — a step towards specific photoaffinity labeling, *Trends Biochem. Sci.,* 7, 310, 1982.
83. **Nathanson, N. M. and Hall, Z. W.,** *In situ* labeling of *Torpedo* and rat muscle acetylcholine receptor by a photoaffinity derivative of α-bungarotoxin, *J. Biol. Chem.,* 255, 1698, 1980.
84. **Erni, B. and Khorana, H. G.,** Fatty acids containing photoactivatable carbene precursors. Synthesis and photochemical properties of 3,3-bis(1,1-difluorohexyl)diazirine and 3-(1,1-difluorooctyl)-3H-diazirine, *J. Am. Chem. Soc.,* 102, 3888, 1980.
85. **Hearne, M. and Benisek, W. F.,** Photoaffinity modification of Δ^5-3-ketosteroid isomerase by light-activatable steroid ketones covalently coupled to agarose beads, *Biochemistry,* 22, 2537, 1983.
86. **Burnett, B. K., Robson, R. J., Takagaki, Y., Radhakrishnan, R., and Khorana, H. G.,** Synthesis of phospholipids containing photoactivatable carbene precursors in the headgroups and their crosslinking with membrane proteins, *Biochim. Biophys. Acta,* 815, 57, 1985.
87. **Gupta, C. M., Costello, C. E., and Khorana, H. G.,** Sites of intermolecular crosslinking of fatty acyl chains in phospholipids carrying a photoactivable carbene precursor, *Proc. Natl. Acad. Sci. U.S.A.,* 76, 3139, 1979.
88. **Brunner, J. and Richards, F. M.,** Analysis of membranes photolabeled with lipid analogues, *J. Biol. Chem.,* 255, 3319, 1980.
89. **Letokhov, V. S.,** Laser biology and medicine, *Nature (London),* 316, 325, 1985.
90. **Yemul, S., Berger, C., Estabrook, A., Edelson, R., and Bayley, H.,** The delivery of phototoxic drugs to selected cells, *Ann. N.Y. Acad. Sci.,* 446, 403, 1985.
91. **Park, C. S., Hillel, Z., and Wu, C.-W.,** Molecular mechanism of promoter selection in gene transcription. I. Development of a rapid mixing-photocrosslinking technique to study the kinetics of *E. coli* RNA polymerase binding to T7 DNA, *J. Biol. Chem.,* 257, 6944, 1982.
92. **Park, C. S., Wu, F. Y.-H., and Wu, C.-W.,** Molecular mechanism of promoter selection in gene transcription. II. Kinetic evidence for promoter search by a one-dimensional diffusion of RNA polymerase molecule along the DNA template, *J. Biol. Chem.,* 257, 6950, 1982.
93. **Karlin, P., Cox, R., Kaldany, R.-R., Lobel, P., and Holtzman, E.,** The arrangement and function of the chains of the acetylcholine-receptor of *Torpedo* electric tissue, *Cold Spring Harbor Symp. Quant. Biol.,* 48, 1, 1983.
94. **Muhn, P., Fahr, A., and Hucho, F.,** Rapid laser flash photoaffinity labeling of binding sites for a noncompetitive inhibitor of the acetylcholine receptor, *Biochemistry,* 23, 2725, 1984.
95. **Nassal, M., Abercrombie, D., and Khorana, H. G.,** unpublished observations.

Chapter 10

TRANSITION METAL CARBONYL COMPLEXES FROM DIAZIRINES

H. Kisch

TABLE OF CONTENTS

I. INTRODUCTION

Transition metal carbonyl complexes of amines and olefins constitute a well-established field in organometallic chemistry; however, complexes of 1,2-diazenes (RN=NR) have been investigated only in the past 15 years.[1] Although diazenes are isoelectronic with olefins, they are expected to combine the coordinating properties of amines and olefins due to the presence of two lone pairs and one π and π^* orbital capable of forming a σ- and π-type complex, respectively. Accordingly, 1,2-diazenes may act as two-, four-, or six-electron donors.[1] In general, *cis*-1,2-diazenes fixed in a cyclic ring are more reactive toward transition metal carbonyls than the *trans*-isomers. The types of complexes obtained depend on the ring size of the 1,2-diazene.

Diazirines were shown to stabilize certain complexes which could not be isolated in the case of larger-ring ligands.[1] This may arise from the reduced steric demand of the three-membered ligand and from electronic effects. Quantum chemical calculations reveal that the lone-pair orbitals are delocalized to a considerable extent onto the neighboring C–N bonds.[2] The highest occupied molecular orbital (HOMO), corresponding to the antisymmetric combination of the two lone-pair orbitals has about 30% C–N and 70% lone-pair character in a six-membered diazene, whereas it exhibits 68 and 32%, respectively, in diazirine.[3] In the latter case, an intensive mixing of the lone pairs with a C-2p orbital results in a HOMO which should be viewed as a bonding σ (C–N) rather than as a nitrogen lone-pair orbital.

Further interest in the coordinating properties of diazirines stems from the problem of biological dinitrogen reduction. The initial step of this process is formulated as side-on binding of N_2 to a molybdenum center.[4] The resulting three-membered metallacycle may be viewed as a molybdodiazirine. In addition, diazirines per se are interesting ligands capable of isomerizing to diazoalkanes or decomposing to carbenes and dinitrogen.

II. REACTION OF DIAZIRINES WITH TITANIUM CARBONYLS

When dicarbonylbis(cyclopentadienyl)titanium is reacted with dimethyldiazirine at room temperature, an unexpected reaction occurs.[5] Instead of a cleavage into dinitrogen and dimethylcarbene derived products, the diazirine reacts by cleavage of the N=N bond (Figure 1).

The formation of the isocyanato group may be rationalized by initial formation of (η^5-$C_5H_5)_2$Ti(CO)(dimethyldiazirine). σ-Donation via the HOMO, which is mainly a bonding C–N orbital, should weaken this bond and facilitate C–N cleavage to (η^5-$C_5H_5)_2$Ti(CO)(N–N=C(CH$_3$)$_2$). Migration of carbon monoxide to the nitrenic nitrogen and concomitant N–N cleavage may eventually afford complex **1**. The characteristic vibration of the isocyanato ligand of **1** is observed at 2200 cm^{-1} in benzene solution.

This reaction may be compared with the formation of isocyanato titanium complexes by cleavage of dinitrogen in the system[6] (η^5-$C_5H_5)_2$TiCl$_2$/Mg/N$_2$/CO$_2$.

III. REACTIONS OF DIAZIRINES WITH GROUP VI CARBONYLS

Dimethyldiazirine and pentamethylenediazirine smoothly displace tetrahydrofuran (THF) or acetonitrile from M(CO)$_5$L at room temperature (Figure 2).[7,8] Shortly after addition of the diazirine, the colorless solution becomes deep red due to the formation of the bridged complex **3**. The yellow mononuclear complexes **2** are isolated by sublimation at room temperature. In solution they decompose to **3** and the free diazirine. This process occurs even in the solid state, although very slowly suggesting a weak metal-diazirine bond. It cannot be completely ruled out that **2** is a π-complex. However, the formulation as a σ-complex is preferred due to the similarity of IR and UV-VIS spectra with the analogous 1,2-diazetine complexes.[9] The ν(CO) absorptions of **2a** (M=W) appear at 2083(m), 1957(vs), 1946(vs), and 1930(vw) cm^{-1}. This is 10 to 20 cm^{-1} higher than in the 1,2-diazetine analog, indicating stronger π-back donation in the diazirine case.

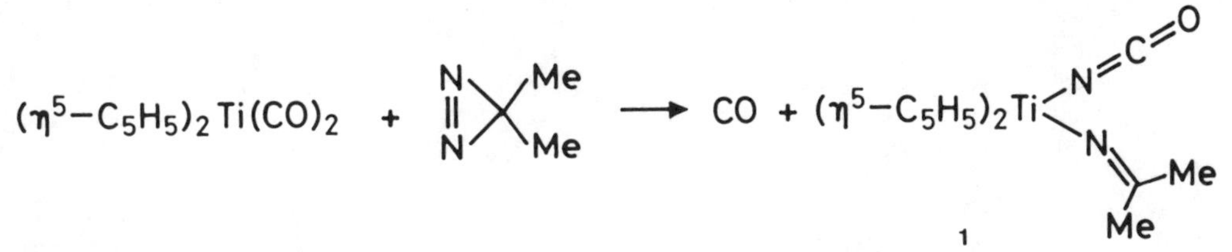

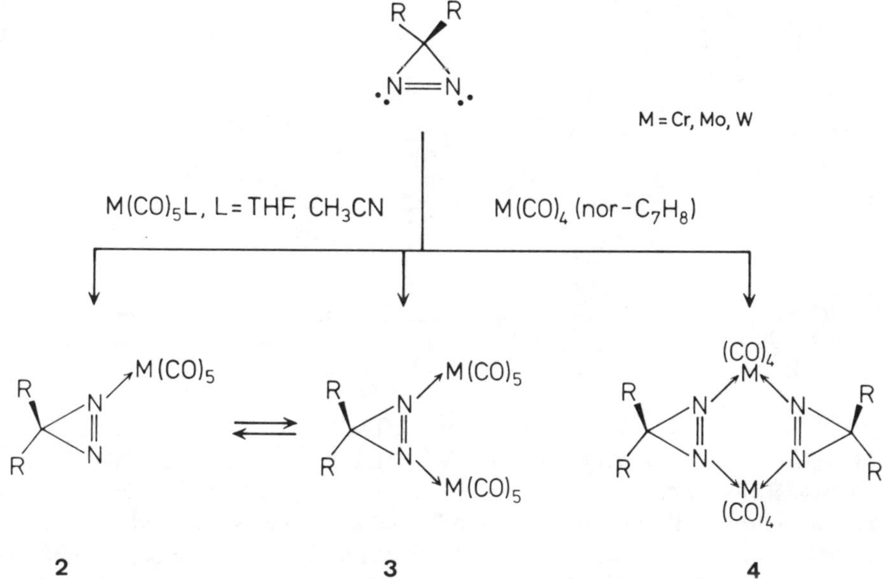

FIGURE 1.   Reaction of dicarbonylbis(cyclopentadienyl)titanium with dimethyldiazirine.

FIGURE 2.   Ligand displacement reactions of dimethyl- and pentamethylenediazirine with group VIb metal carbonyls.

Complexes **3** containing a single bridging diazene group are obtained in good yields when a twofold excess of the metal carbonyl is used. The appearance of five ν(CO) bands in the IR spectrum at 2068(m), 1975(vs), 1958(vs)(sh), 1954(vs), and 1945(m) cm^{-1} suggests that free rotation around the M–N bond is restricted and that the M(CO)$_5$ fragment lacks C$_{4v}$ local symmetry. This may be predominantly caused by steric interference of the two M(CO)$_5$ groups, since the analogous *trans*-diimine complex exhibits only three ν(CO) bands in accordance with a local C$_{4v}$ symmetry.[10] The structure of **3** is based on comparison with heterodimetal analogs which have been investigated via X-ray analysis *(vide infra)*.

When (norbornadiene) M(CO)$_4$ reacts with diazirines at room temperature, double bridged complexes **4** of metallic-blue color are formed. The dimetallacycle is planar with Mo–N and N–N bond lengths of 2.13 and 1.26 Å, respectively, as determined for **4b** (M=Mo).[7] In the free ligand the N–N distance is 1.24 Å.[11] This suggests some π-back donation into the π^* (N=N) orbital. The solution IR spectra of these complexes in the range from 2010 to 1910 cm^{-1} usually exhibit four strong ν(CO) bands in agreement with a D$_{2d}$ symmetry.

UV-VIS absorption spectra of the tungsten complexes **2b**, **3b**, and **4b** are shown in Figure 3. The low-energy maximum of **2b** (M=W) at 25,400 cm^{-1} (394 nm) contains a shoulder at 28,400 cm^{-1} (352 nm), indicating the presence of at least two absorption bands. By analogy with pentacarbonyltungsten complexes of six-membered diazenes,[12] the high-energy shoulder is tentatively assigned to metal centered (MC) transitions, the lower-energy max-

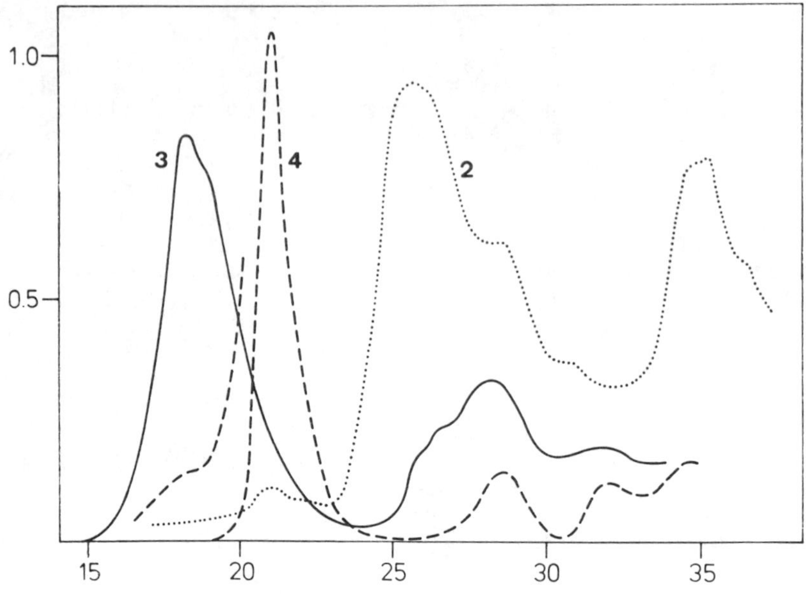

FIGURE 3.   UV-VIS spectra of the tungsten pentamethylenediazirine complexes **2b** (. . .), **3b** (——), and **4b** (----) in 2,2-dimethylbutane/n-pentane = 8/3 at $-125°C$. Vertical axis: absorbance. Horizontal axis: wave number $\times$ 10^{-3} (cm^{-1}).

imum to a charge-transfer metal $\to$ diazirine (CTML) transition. In the single bridged complex **3b** (M=W) at room temperature the MC bands are located at about the same energy levels whereas the other band is shifted to 18,300 cm^{-1} (ϵ = 25,000 M^{-1} cm^{-1}). The latter may again be of the CTML type, but the empty ligand π^*-orbital is expected to be delocalized onto the metal carbonyl groups. Substituting tungsten for molybdenum or chromium results in a shift to 19.2 and 18.9 $\times$ 10^3 cm^{-1}, respectively. The spectra are typical for a M(RN=NR)M arrangement irrespective of a *cis-* or *trans*-configuration for the bridging diazene. Accordingly, the spectra are almost superimposable with those of the corresponding *trans*-diimine complexes.[13] Upon cooling solutions of **3b** to $-125°C$ shoulders appear on both major bands (Figure 3) because of a conformational equilibrium (Figure 4). Rotation around the M–N bond of one half of the complex molecule creates two conformations where the N=N bond is eclipsed (e) or staggered (s) with respect to one of the four equatorial M–CO groups. The complete complex molecule should then exist in the three conformers (s,s), (e,e), and (s,e).

Steric pressure between the carbonyl groups should be least significant in the conformer (s,e), as indicated by the Newman projections in Figure 4. In fact, X-ray analysis of the heterodimetal complex **5** (ML$_n$=Cr(CO)$_5$, M′L$_m$=Mo(CO)$_5$) reveals this conformation.[14] The projection of the N=N bond onto the planes of the equatorial CO groups is aligned along Cr–C(7) , while it bisects the Mo–C(5) and Mo–C(4) bonds. Both metal atoms are displaced only by 0.15 Å from the diazirine plane. The N–N distance of 1.26 Å is the same as observed in **4b** (M=Mo)[7] (Figure 5).

Complexes of type **5** are prepared from complexes of type **2** and metal carbonyl fragments containing readily displaceable ligands such as THF (Figure 6). The heterodimetal complexes obtained (**5**) may be relevant to the mechanism of N$_2$ reduction in the enzyme nitrogenase. It was proposed that diimine may be reduced via a bridged Fe/Mo complex,[15] and therefore **5** may serve as a model for this intermediate. While complexes **5** were isolated with the metal pairs Cr/Mo, Cr/Fe, W/Fe, and Cr/Mn, the Fe/Mo complex was too reactive to be isolated.[14] IR and UV-VIS spectra resemble those of the homodimetal complexes (**3**).

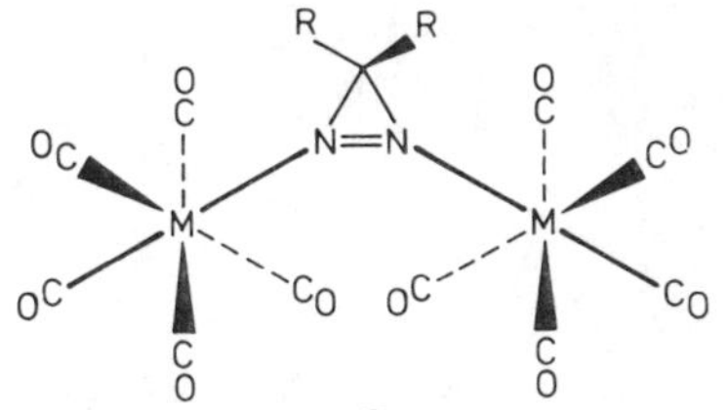

FIGURE 4.   Conformers of single bridged diazirine complexes of type 3.

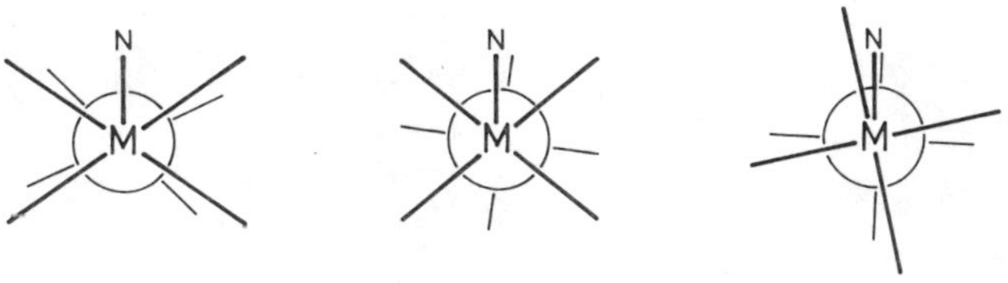

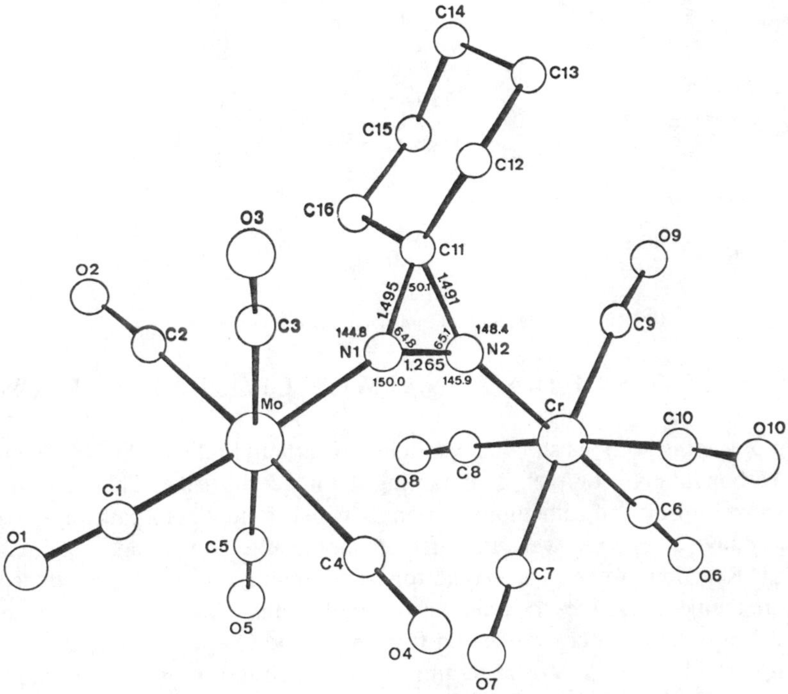

FIGURE 5.   Molecular structure of (pentamethylenediazirine)CrMo(CO)$_{10}$.

The UV-VIS spectra of the dimetallacycles **4** closely resemble those of **3** (Figure 3).[8] Only the low-energy band is shifted hypsochromically to 21,000 cm^{-1}. Unlike **3**, however, the spectrum is not temperature dependent because a similar conformational equilibrium is impossible in this case. Accordingly, the half width of the absorption band in **4** is only about 50% of that in **3**. This again supports the presence of conformers with slightly different spectra in the case of **3**.

In addition to the complexes discussed, pentacarbonylmetal carbene complexes are obtained in low yields from 3-methoxy-3-phenyldiazirine and the hexacarbonylmetal upon irradiation.[16]

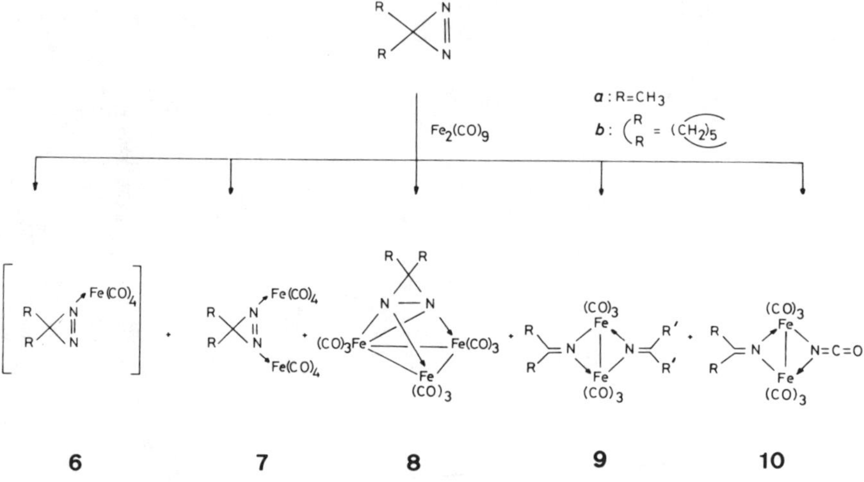

MLn | Cr(CO)₅ | Cr(CO)₅ | W(CO)₅ | Cr(CO)₅
M'Lm | Mo(CO)₅ | Fe(CO)₄ | Fe(CO)₄ | Mn(CO)₂(η⁵-MeC₅H₄)

FIGURE 6.   Ligand displacement reactions of (diazirine)M(CO)₅ complexes.

FIGURE 7.   Reaction of enneacarbonyldiiron with diazirines.

IV. REACTIONS OF DIAZIRINES WITH GROUP VIII CARBONYLS

The two preceding sections showed that titanium carbonyls cleave the N=N bond, whereas group VI carbonyls give complexes containing the intact diazirine. This section focuses on the reactions with iron and ruthenium carbonyls which follow both pathways. Historically, the iron carbonyl complexes were the first organometallic compounds obtained from diazirines.[7,17,18] Reaction of enneacarbonyldiiron with dimethyl- or pentamethylenediazirine in THF at room temperature gives the relatively unstable complexes of type **7**. They are obtained as deep blue crystals upon crystallization from n-hexane at $-70°C$. The structure is based on spectral data like the UV-VIS spectrum of **7a**, which has a low-energy band at $\bar{\nu}_{max} = 18,300\,cm^{-1}$ ($\epsilon = 3500\,M^{-1}\,cm^{-1}$). This resembles the spectrum of the analogous complexes of group VI carbonyls previously discussed. From the appearance of six $\nu(CO)$ bands in the solution IR spectrum, the absence of free rotation around the Fe–N bond may be inferred, again paralleling the properties of the group VI compounds **3**. IR reaction spectroscopy reveals formation of an intermediate tetracarbonyliron complex (**6**), which was not isolated. The complexes of type **7** in solution slowly decompose into the stable compounds **9** and **10** (Figure 7). These can be directly synthesized by conducting the reaction in n-hexane or toluene. However, the reaction is much slower than in THF, and gives only traces of **7** and small amounts of the cluster **8** which is preferentially formed in the case of five-membered diazenes.[19]

The structures of **9** and **10** have been solved by X-ray analysis.[18,20] Important features are summarized in Figure 8. They belong to the well-established $L_2Fe_2(CO)_6$ structure, where

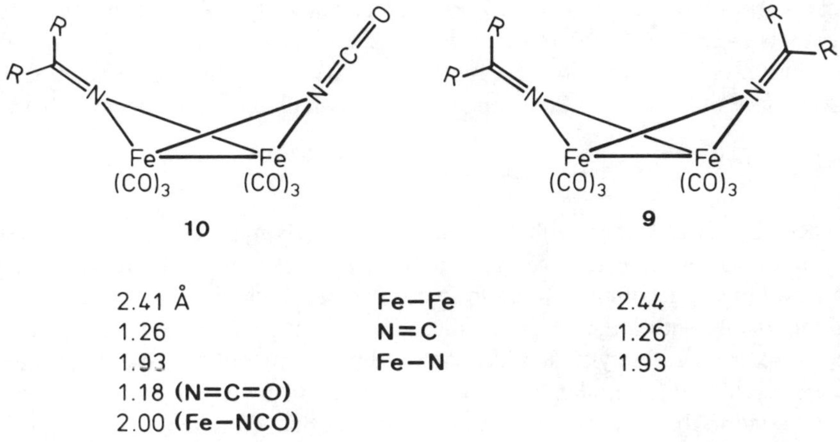

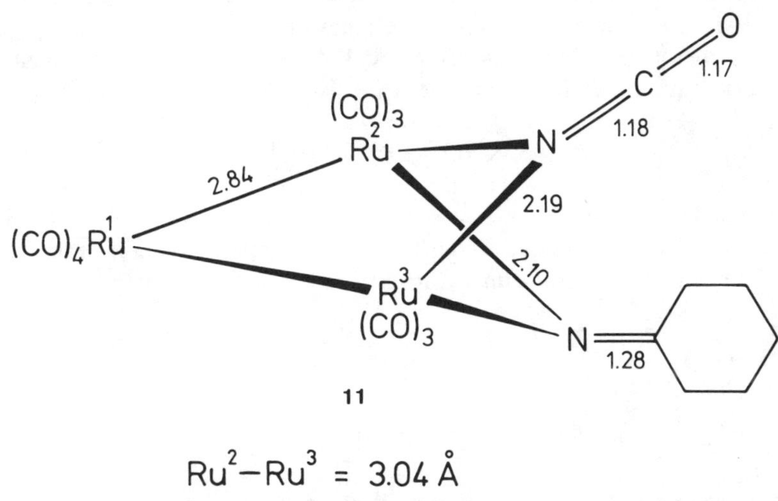

2.41 Å	**Fe—Fe**	2.44
1.26	**N=C**	1.26
1.93	**Fe—N**	1.93
1.18 **(N=C=O)**		
2.00 **(Fe—NCO)**		

FIGURE 8.   Selected bond distances in complexes **9** and **10**, R–R=pentamethylene.

$$Ru^2\!-\!Ru^3 = 3.04\ \text{Å}$$

FIGURE 9.   Selected bond distances in complex **11**.

L is a bridging three-electron donor. The dihedral angle between the two Fe_2N planes is 109.5 and 119.8° in **9** and **10**, respectively. Complex **9** was also synthesized from R_2C=NLi and $Fe(CO)_4I_2$ or from $Na_2Fe(CO)_4$ and 2-bromo-2-nitrosopropane in low yields.[21,22]

The formation of **9** in our system does not occur via an intermediate azine which may be formed from the diazirine and its carbene. A possible rationalization is initial formation of **7** which is attacked by a second molecule of diazirine to give a $(diazirine)_2Fe_2(CO)_8$ intermediate similar to **4**. This could rearrange to **9** by cleavage of two C–N bonds, migration of two molecules CO to two nitrenic nitrogen atoms with concomitant N–N cleavage and elimination of two isocyanato groups. This is supported by the additional formation of mixed complexes **9** (R ≠ R′) when **7a** is reacted with pentamethylenediazirine and the isolation of an inorganic isocyanate.[17] The formation of **10** may be rationalized in a similar manner. However, as in the case of **9**, the actual mechanism is presently unknown.

This unusual cleavage of the diazirine N=N bond (Figure 9) occurs also with ruthenium carbonyls.[23] Heating dodecacarbonyltriruthenium with pentamethylenediazirine yields a yellow, trinuclear complex of structure **11**. This resembles the iron complex **10**, except that the two tricarbonylmetal groups are bridged by a tetracarbonylmetal group in **11**. The interplanar angle between the two Ru(2)Ru(3)N planes is 124.5°, and the distance of the nitrogen atoms is 2.67 Å. If the bridging isocyanato and ketiminato groups are considered as three-electron donors, it is not necessary to assume a Ru^2-Ru^3 single bond. The corresponding distance of 3.04 Å is 0.20 Å longer than the two metal-metal bonds of the molecule.

V. SUMMARY

The reactions of diazirines with transition metal carbonyls is a relatively new field, revealing a rich and unexpected chemistry. The complexes with intact diazirines such as **3** reflect their ability to accommodate up to two σ-bonded metal carbonyl groups, a capacity which is impossible with larger ring-size diazenes. Although these complexes are stable in the case of group VI carbonyls, they decompose by cleavage of the N=N bond in the case of iron carbonyls. The dissimilar behavior may be due to the stronger tendency to form a metal-metal bond in the latter case. While this can be achieved in the case of larger diazene rings without N=N cleavage, it is impossible to accomplish because of the larger metal-metal distance in the diazirine complexes. The cleavage with dicarbonyl-bis(cyclopentadienyl)titanium may occur as a result of the strong reducing property of this compound. The dimetallacycles of type **4** are noteworthy for stability and the presence of energetically well separated MC and CT excited states making them excellent candidates for the study of wavelength-dependent photochemistry.

ACKNOWLEDGMENTS

I am deeply indebted to my former co-workers Prof. Dr. A. Albini, Dr. G. Avar, Dr. R. Battaglia, and Dr. P. Mastropasqua who conducted research in this field at the Max-Planck-Institut für Strahlenchemie, Mülheim a.d.Ruhr.

REFERENCES

1. **Albini, A. and Kisch, H.,** Complexation and activation of diazenes and diazo compounds by transition metals, *Top. Curr. Chem.,* 65, 105, 1976.
2. **Hoffmann, R.,** Excited states and photochemistry of diazirines and diazomethanes, *Tetrahedron,* 22, 539, 1966; **Haselbach, E., Heilbronner, E., Mannschreck, A., and Seitz, W.,** Interaction of lone-pairs in 3,3-dimethyldiazirine, *Angew. Chem.,* 82, 879, 1970; *Angew. Chem. Int. Ed.,* 9, 902, 1970.
3. **Olbrich, G.,** private communication, 1981.
4. **Newton, W., Postgate, I. R., and Rodriguez-Barrueco, C., Eds.,** *Recent Developments in Nitrogen Fixation,* Academic Press, New York, 1977; **Liu, N. H., Strampach, N., Palmer, J. G., and Schrauzer, G. N.,** Reduction of molecular nitrogen in molybdenum(III-V)hydroxide/titanium(III)hydroxide systems, *Inorg. Chem.,* 23, 2772, 1984.
5. **Kisch, H. and Avar, G.,** unpublished results, 1978.
6. **Sobota, P., Jezowska-Trzebiatowska, B., and Janas, Z.,** Insertion of carbon dioxide into the metal-nitrogen bond formed in the reaction with molecular nitrogen, *J. Organomet. Chem.,* 118, 253, 1976.
7. **Beck, W. and Danzer, W.,** Reactions of metal carbonyls with diazirines, *Z. Naturforsch.,* 30b, 716, 1975.
8. **Battaglia, R., Matthäus, H., and Kisch, H.,** Reactions of diazirines with chromium, molybdenum and tungsten carbonyls, *J. Organomet. Chem.,* 193, 57, 1980.
9. **Albini, A. and Kisch, H.,** Reactions of 1,2-diazetines and 1,2-diazetine-1-oxides with chromium and tungsten carbonyls, *Z. Naturforsch.,* 37b, 468, 1981.
10. **Sellmann, D., Brandl, A., and Endell, R.,** The structure of diimine: IR spectra of N_2H_2-, N_2D_2- and $^{15}N_2H_2[Cr(CO)_5]_2$, *Angew. Chem.,* 85, 1122, 1973; *Angew. Chem. Int. Ed.,* 12, 1019, 1973; **Huttner, G., Gartzke, W., and Allinger, K.,** Structure of diimine: x-ray analysis of $N_2H_2[Cr(CO)_5]_2$. 2 THF, *Angew. Chem.,* 86, 860, 1974; *Angew. Chem. Int. Ed.,* 13, 822, 1974.
11. **Wollrab, J. E., Scharpen, L., LeRoy, H., and Ames, D. P.,** Rotational spectra of substituted diazirines. I. Dimethyldiazirine, *J. Chem. Phys.,* 49, 2405, 1968.

12. **Frazier, C. C. and Kisch, H.,** Nature of the lowest excited states and dynamic behavior of group VIb (diazene)pentacarbonylmetal complexes, *Inorg. Chem.,* 17, 2736, 1978.

13. **Sellmann, D., Brandl, A., and Endell, R.,** Diimine, hydrazine, and ammonia complexes of molybdenum, *J. Organomet. Chem.,* 97, 229, 1975.

14. **Battaglia, R., Kisch, H., Krüger, C., and Liu, L.-K.,** Synthesis and properties of heterodimetal complexes $L_nM\text{-}(cis\text{-}RN{=}NR)M'L_m$, *Z. Naturforsch.,* 35b, 719, 1980.

15. **Stiefel, E. I.,** The mechanism of nitrogen fixation, in *Recent Developments in Nitrogen Fixation,* Newton, W., Postgate, I. R., and Rodriguez-Barrueco, C., Eds., Academic Press, New York, 1977, 69.

16. **Chaloner, P. A., Glick, G. D., and Moss, R. A.,** Synthesis of metal carbene complexes from diazirines, *J. Chem. Soc. Chem. Commun.,* 880, 1983.

17. **Albini, A. and Kisch, H.,** Complexation and cleavage of the N=N bond of diazirines by iron carbonyls, *J. Organomet. Chem.,* 94, 75, 1975.

18. **Tikhonova, I. A., Enisk, K., Andrianov, V. G., Byalotskii, S. F., Struchkov, Yu. T., Shur, V. B., Shmitts, E., and Volpin, M. E.,** Reaction of 3,3-pentamethylenediazirine with nonacarbonyldiiron and aquopentaamine derivatives of divalent ruthenium, *Sov. J. Koord. Khim.,* 2, 1274, 1976.

19. **Krüger, C., Trautwein, A., and Kisch, H.,** Structure and Mößbauer data of (2,3-diazanorbornene)Fe$_3$(CO)$_9$, *Z. Naturforsch.,* 36b, 205, 1981.

20. **Krüger, C.,** private communication, 1974.

21. **King, R. B. and Douglas, W. M.,** Organonitrogen derivatives of metal carbonyls. VI. Novel products from reactions of 2-bromo-2-nitrosopropane with metal carbonyl anions, *Inorg. Chem.,* 13, 1339, 1974.

22. **Kilner, M. and Midcalf, C.,** Organonitrogen groups in metal carbonyl complexes. VII. Methylene-amino-derivatives of iron carbonyls, *J. Chem. Soc. Dalton,* 1974, 1620.

23. **Mastropasqua, P., Riemer, A., Kisch, H., and Krüger, C.,** Cleavage of the N=N bond of diazirines by Ru$_3$(CO)$_{12}$, *J. Organomet. Chem.,* 148, C40, 1978.

Chapter 11

ELECTROCHEMISTRY OF DIAZIRINES AND DIAZIRIDINES

Clive M. Elson and Michael T. H. Liu

I. INTRODUCTION

In this chapter, a review of electrochemical studies of diazirines and diaziridines undertaken before 1980 will be presented along with some recent developments. The application of two electrochemical techniques, cyclic voltammetry and double potential step chronoamperometry, to the determination of rate constants of chemical reactions which succeed the electron transfer step will also be described.

Since only a brief and simplified introduction to the technique of cyclic voltammetry will be offered here, the reader should consult the texts and monographs[1-3] on electrochemical techniques for more details. Reviews of diazirines and diaziridines prior to 1982 may be found as part of the general reviews of the electrochemistry of heterocyclic or organic compounds.[4-6] Pre-1982 work focused on the reduction of the nitrogen-nitrogen double bond of diazirines or the single bond of diaziridines, and on the oxidation of diaziridines to diazirines, i.e., electrode processes that occurred in protic media or involved hydrogen ion transfers. More recently, the electrochemical behavior of a series of chlorophenyl- and alkylphenyl-diazirines in aprotic media[7,8] has been studied with the result that a new class of nitrogen-containing radicals, the diazirine anion radicals, have been identified. This later work will be discussed in some detail.

II. ELECTROCHEMICAL METHODS

A. Cyclic Voltammetry

Cyclic voltammetry is an electrochemical technique in which the potential applied to a working electrode (site of reaction of interest) is controlled while the resultant current is monitored. The "cyclic" term refers to the variation of the applied potential. Starting from a chosen initial value, the applied potential is altered (swept) at a constant rate toward either more negative or more positive potentials until a preset limit is reached; at this point, the direction of potential scan is reversed and the applied potential returns to the initial value. The control of the applied potential and potential scan rate is achieved with a combination of a potentiostat and waveform generator. The electrolysis current is usually displayed on a recorder or oscilloscope.

For this technique, three electrodes are connected to the potentiostat: the working electrode, a reference electrode, and the counter electrode. The potential of the reference electrode is constant and is determined by the construction of the electrode and its environment. Specific details concerning the construction of reference electrodes may be found in Reference 1. One of the simplest reference electrodes to construct involves dissolving silver nitrate (0.1 M) in a portion of the solvent of choice and placing a clean silver wire in the solution. The silver/silver nitrate electrode is then separated from the sample solution by a Vycor™ plug sealed in a piece of heat-shrink tubing.

The applied potential appears as a potential difference between the reference electrode and the working electrode (the current never flows through the reference electrode). The choice of the working electrode is based on the following criteria: (1) surface characteristics can be reproduced with minimal electrode pretreatment; (2) absorption of the reactants and products of the electrode reaction is minimized; (3) the material is noncatalytic with respect to the oxidation or reduction of the solvent system; (4) the material is a good electrical conductor; and (5) where possible, a planar surface may be formed which simplifies the treatment of the data. The most commonly employed electrodes are made of either platinum, vitreous (glassy) carbon, graphite, or mercury. A piece of vitreous carbon heat sealed into glass tubing and polished with jewelers' rouge provides a good general purpose electrode for most organic solvents. The electrolysis current associated with the reduction or oxidation of the electroactive species flows between the working and counter electrodes. The counter

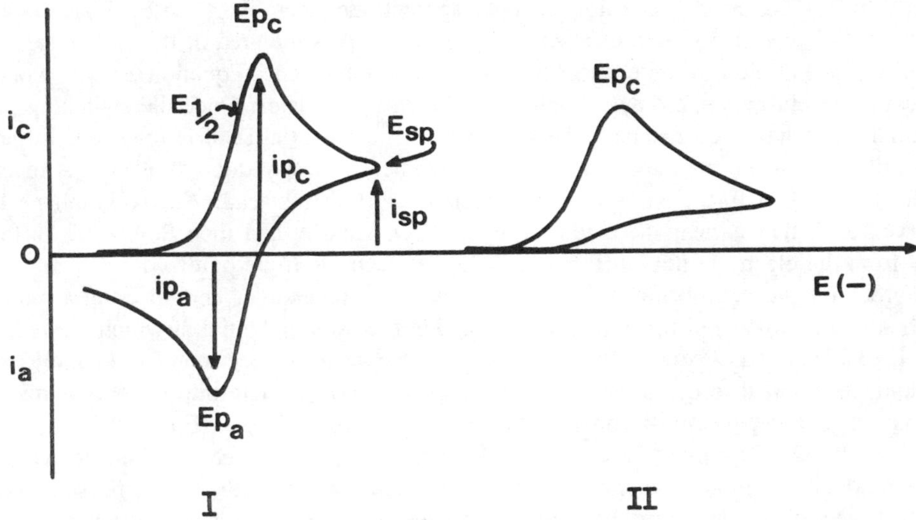

FIGURE 1.   Cyclic voltammograms: i = current, E = potential, c = cathodic (reduction), a = anodic (oxidation), sp = switching point, p = peak, and $E_{1/2}$ = polarographic half wave potential.

electrode is often a piece of platinum wire or a platinum gauze (when currents exceed a few microamperes).

The choice of solvent is usually dictated by the solubility of the sample. Nevertheless, sufficient supporting electrolyte (inert salt, such as a tetraalkylammonium halide) must be dissolved in the solvent to produce a 0.1 M solution. The electrolyte is required to carry the current between the counter and working electrolytes and to minimize the ohmic resistance of the solution. This latter point is important, because a substantial ohmic resistance will distort the shape of cyclic voltammograms, especially at high potential scan rates, which in turn precludes any interpretation of the data. Details of cells, solvents, and electrolytes may be found in the reference texts.

Two typical cyclic voltammograms (plots of observed current against applied electrode potential) are shown in Figure 1. In both cases, a solution species is being reduced on the forward scan (from left to right) to more negative (more reducing) potentials. If a solution species were being oxidized, then the cyclic voltammograms would appear as the inverse of those of Figure 1.

Case I of Figure 1 is designated as a "reversible" system:

$$O + ne^- \rightleftharpoons R \tag{1}$$

where O and R are the oxidized and reduced forms of the compound respectively, and n is the number of electrons transferred. During the forward scan, at some point the potential is sufficiently reducing to convert O to R and a current appears. The current increases sharply with small changes in applied potential as O near the electrode surface is reduced. Once all the O near the electrode has been reduced, i.e., a depletion layer has been formed, then the current reaches a maximum and begins to decay, even though the potential is still being varied. The current is now dependent upon the rate of diffusion of O across the depletion layer to the surface of the electrode and is given by:

$$i = \frac{nFD^{1/2}AC^*}{\pi^{1/2}t^{1/2}} \tag{2}$$

under diffusion-controlled conditions to a planar electrode where $E_{applied} > E_p$, F is Faraday's Constant, D is the diffusion coefficient of the species, A is the area of the electrode, C* is the bulk concentration of the electroactive species, and t is time. Equation 2 is the working expression of chronoamperometry which will be discussed later. Once the switching point potential, Esp, has been reached, the direction of the potential scan is reversed. Initially, O is still reduced, but as the potential moves anodically (less reducing) a point is reached where the reduced form, R, is oxidized back to starting material. An oxidation peak is observed as R that is near the electrode is oxidized initially and then R that has diffused away from the electrode now diffuses back to the electrode to be oxidized.

In order for the equilibrium (1) to be designated as reversible, certain criteria must be satisfied: (1) the oxidized form must be reduced at the same potential at which the reduced form is oxidized and (2) the electron transfer rates between the electrode and O and R must be rapid. As applied to cyclic voltammetry, these criteria translate into the following conditions: Ep_c is independent of concentration and scan rate, v; $Ep_c - Ep_a = 59/n$ mV; and $Ep_c - E_{1/2} = 28.5/n$ mV at 25°C. The peak current, ip_c, will increase as both the concentration and the potential scan rate are increased. The condition that must be satisfied to demonstrate a diffusion controlled electrode process is $ip_c/v^{1/2}$ equals a constant.

Once it has been shown that the reduction or oxidation of a compound is reversible and diffusion controlled, then the product of the electrode reaction can be characterized as having the same chemical composition and essentially the same structure (allowing for small changes in bond lengths and degree of solvation) as the parent compound and differing only in the number of electrons. This conclusion is based on the concept of reversibility: the oxidized and reduced forms undergo electron transfers at the same electrode potentials. Hence, the energies of the redox molecular orbitals that accept the electrons in the reduction step or lose electrons during oxidation must be equal. This condition is usually only satisfied when the contributions of the various atomic orbitals to the redox molecular orbitals in the reduced and oxidized forms are the same, i.e., neither the chemical composition nor the geometry is altered by the electron transfer process.

Case II of Figure 1 represents an "irreversible" system:

$$O + ne^- \rightarrow R \tag{3}$$

Two features distinguish it form the reversible case: (1) the rising part of the curve is drawn out and (2) the oxidation wave on the reverse (right to left) scan is absent. The rising part of the reduction wave is drawn out because the rate of the electrode process is "slow", i.e., there is an energy barrier associated with the transfer of the electron to the solution species and a characteristic rate constant can be measured. Now, as the scan rate is increased, it is possible to "outrace" the rate of electron transfer and, hence, the peak appears at more negative potentials, i.e., Ep_c shifts cathodically as v increases. Usually, the irreversible nature of the electrode reaction is due to some chemical change that occurs concurrently with the charge transfer, e.g., the two-electron reduction of dichloroacetate is accompanied by the release of chloride ion and uptake of hydrogen ion[1]:

$$Cl_2HCCOO^- + e^- \rightarrow Cl H\dot{C}COO^- + Cl^- \tag{4}$$

$$Cl H\dot{C}COO^- + H^+ + e^- \rightarrow Cl H_2 CCOO^- \tag{5}$$

In such cases, the energies of the redox molecular orbitals of the reduced and oxidized forms are dramatically different, which explains the irreversibility. Nevertheless, by recording cyclic voltammograms under different solution conditions, qualitative details of the electrode process can be established. In the above example, varying the pH of the electrolyte did not alter the reduction potential of dichloroacetate, which meant that Reaction 4 was the rate-determining electron transfer step. Before readings of current or potential are taken from voltammograms of a sample, the solvent/electrolyte system must be studied under the same conditions to provide baseline corrections.

B. Application of Cyclic Voltammetry to Measurement of Chemical Rate Constants

A complete discussion of chemical reactions that precede and succeed electrode reactions may be found in Reference 9. The present discussion is confined to studying the fairly common case of a first-order or pseudo-first-order chemical reaction occurring subsequently to a diffusion-controlled, reversible electrode process. Furthermore, the working electrode will be considered to be planar and free from absorption of chemical species. The overall process can be represented by Equations 1 and 6:

$$R \xrightarrow{k} Z \tag{6}$$

where Z is an electroinactive product and k is the rate constant of the reaction. In the event that Reaction 6 is very slow, one would expect the cyclic voltammogram to appear as Case I of Figure 1. Moreover, if Reaction 6 were very rapid and the potential scan rate slow, the expected voltammogram would appear something like Case II: because the scan rate is slow, sufficient time elapses as the applied potential is swept from the region where reduction occurs to where R is reoxidized that all of R is converted to Z; hence, no anodic curve is observed. If the scan rate is increased now, one would expect to see voltammograms intermediate between Cases I and II. In fact, there should be some relationship between the rate of the following chemical reaction (Equation 6), the ratio ip_a/ip_c, and the scan rate. This relationship has been established by Nicholson and Shain.[9] Table XI of Reference 9 presents the data required to construct a working curve to calculate k. The data are in the form of i_a/i_c vs. $k\tau$ where i_a/i_c can be calculated from the semiempirical relationship reported by Nicholson[10]:

$$\frac{i_a}{i_c} = \frac{ip_a}{ip_c} + \frac{(0.485)(i_{sp})}{ip_c} + 0.086 \tag{7}$$

referring to Case I of Figure 1 and τ being the time in seconds that it takes to scan from $E_{1/2}$ to E_{sp} (for the reversible case, $E_{1/2}$ occurs at a point 85.17% of the way up the wave or $28.5/n$ mV before Ep_c). The restrictions on the use of this approach are (1) that $k\tau$ should fall in the range 0.05 to 1.25 and (2) that the switching potential should be more than $60/n$ mV beyond Ep_c.

In practice, cyclic voltammograms are recorded at a given solution temperature for a series of potential scan rates and a series of different switching potentials. Values of i_a/i_c and τ are calculated and referred to the working curve constructed from Nicholson and Shain's data to obtain the values of k. To elucidate the mechanism of Reaction 6, kinetic measurements can be made at different temperatures in different solvent/electrolyte systems, and in the presence of an excess of different reagents.

C. Determination of Chemical Rate Constants by Double Potential Step Chronoamperometry

Double potential step chronoamperometry (DPSC) can also be employed to measure first-order or pseudo-first-order rate constants of reactions that follow the charge transfer step. DPSC has a wider application than the cyclic voltammetric technique discussed above because

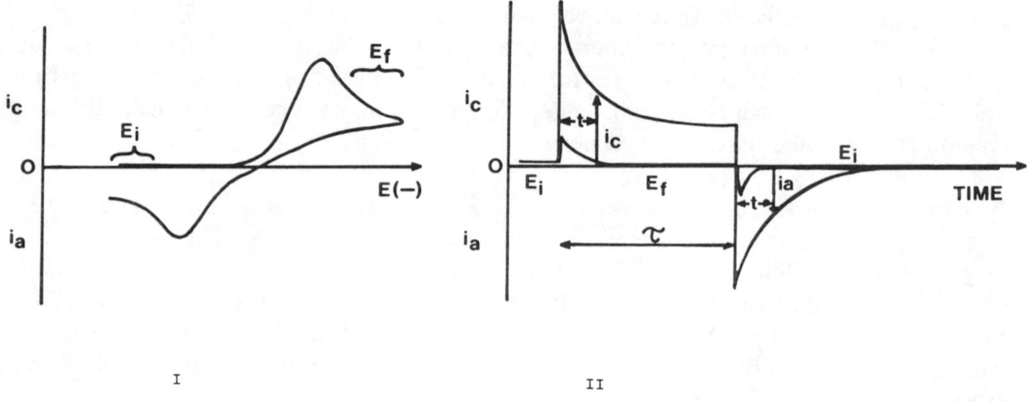

FIGURE 2.   Cyclic voltammogram (I) and double potential step chronoamperogram (II) for O + ne$^-$ $\rightleftharpoons$ R $\xrightarrow{k}$ Z. E_i = initial potential where R $-$ ne$^-$ $\rightarrow$ O, E_f = final potential where O + ne$^-$ $\rightarrow$ R, t = time from application of potential to current reading, and τ = total time of application of E_f.

the restriction of reversibility of the electrochemical reaction is relaxed and now only a gross reversibility is required, "i.e. some conditions must be available where R can be converted back to O".[11] The other restrictions discussed earlier still apply, and the processes are again represented by Equations 1 and 6.

A cyclic voltammogram of the species of interest is usually recorded to determine the potentials where O is reduced (E_f) and R is oxidized (E_i) under diffusion-controlled conditions (Figure 2.I). In DPSC, the potential of the working electrode is again controlled relative to a reference electrode but now it is "stepped" between E_i and E_f, and the resultant current is recorded on a time-based device (e.g., recorder) (Figure 2.II). Initially, the applied potential, E_i is chosen, so that no electrode reaction occurs, but the potential is in the diffusion-controlled region for the oxidation of R. The potential is then instantly stepped to E_f where O is reduced to R and the current is recorded. After a preselected interval, τ, the potential is stepped back to E_i and the reduced material, R, which has not undergone a chemical reaction, is oxidized back to O. The solvent/electrolyte system is studied in an identical manner and the results are applied as a correction. Under diffusion-controlled conditions, the cathodic current-time curve will obey the expression for chronoamperometry (Equation 2). The switching time, τ, is chosen to be of the same magnitude as the half-life of the chemical reaction. Values of i_a/i_c for 4 to 5 different times and a given τ are calculated and compared to working curves presented in Reference 11. This produces $k\tau$ which is then divided by τ to yield the rate constant of the chemical reaction. Two or three different values of τ are usually studied and the current readings are recorded at values of t/τ that correspond to the working curves. An example of the application of DPSC to a kinetic problem is also given in the paper by Schwarz and Shain.[11]

III. ELECTROCHEMICAL BEHAVIOR OF DIAZIRINES AND DIAZIRIDINES

A. Review of Pre-1980 Developments

The first reports of the polarographic behavior of diazirines and diaziridines such as 3,3-pentamethylenediazirine[12] (**1**) and methylethyldiaziridine[13] (**2**) demonstrated that the N=N bond could be reduced or oxidized in aqueous media. The electrode reactions, however, were influenced by the pH of the medium.

In alkaline solution the diazirine (**1**) undergoes a two-electron reduction to yield the corresponding diaziridine (**3**); the $E_{1/2}$ is -1.30 V (vs. SCE) at a pH of 10 in 20% aqueous ethanol.[12] The diaziridine (**3**) can be oxidized at $E_{1/2} = -0.08$ V (vs. SCE) back to the diazirine. The couple is, therefore, only reversible in the gross sense because the difference in reduction and oxidation potentials is about 1.2 V. Lund[12] also reported that diaziridines substituted at one or both nitrogen centers could not be oxidized.

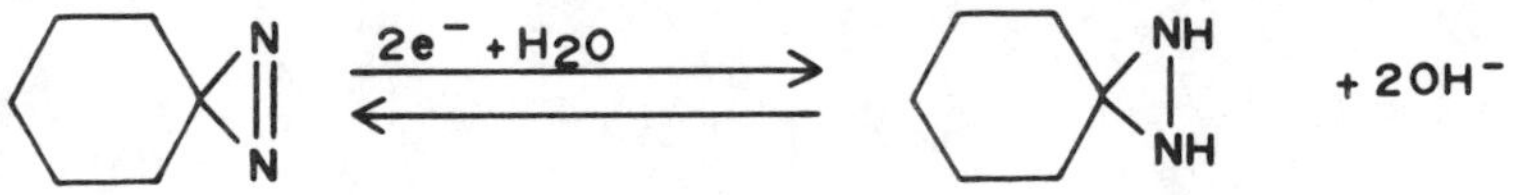

In acidic solution, Kitajev and Budnikov[13] found that prior to the electron transfer step diaziridine (**2**) was protonated. They proposed the mechanism of reduction as:

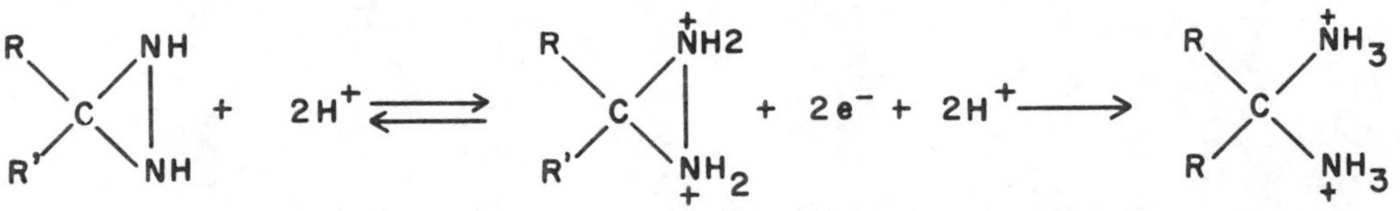

where the resultant diamino species was hydrolyzed to yield ammonia. The half wave potential was approximately -0.5 V at a pH of 3 (vs. AgCl electrode) in 20% aqueous methanol. The diaziridine was not reducible in neutral or alkaline medium. These observations were confirmed by Lund,[12] who reported that in acid solution the diazirine (**1**) underwent a four-electron reduction at pH 3 in 25% alcohol to yield cyclohexanone and ammonia. Researchers also discovered that the diaziridine (**3**) was easier to reduce in acid than the parent diazirine and, hence, the mechanism of diazirine reduction involved the diaziridine (**3**) as an intermediate.[4] The overall process is represented:

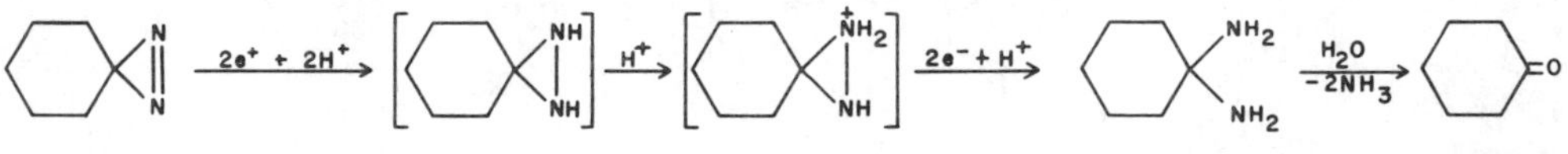

which is a modification of the one discussed above.

In the only other report that appeared during this time, the reduction of di-tert-butyldiaziridinone (**4**) was studied both chemically and electrochemically.[14] In dimethylsulfoxide or dimethylformanide, the diaziridine (**4**) undergoes a diffusion controlled, irreversible, two-electron reduction at -0.76 V (vs. Cd amalgam); the final product of electrolysis in the absence or presence of the proton source, phenol, is di-tert-butylurea:

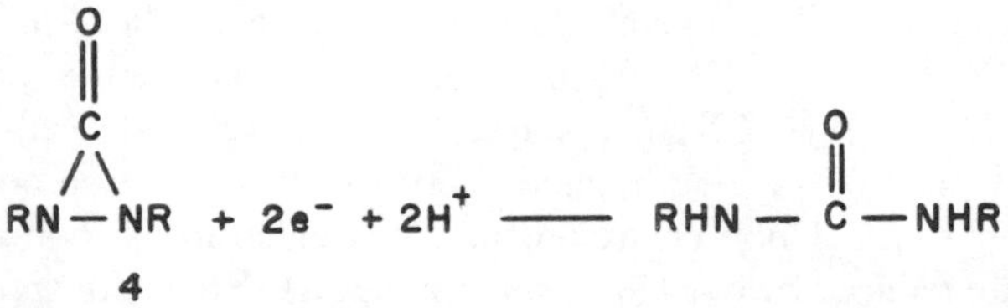

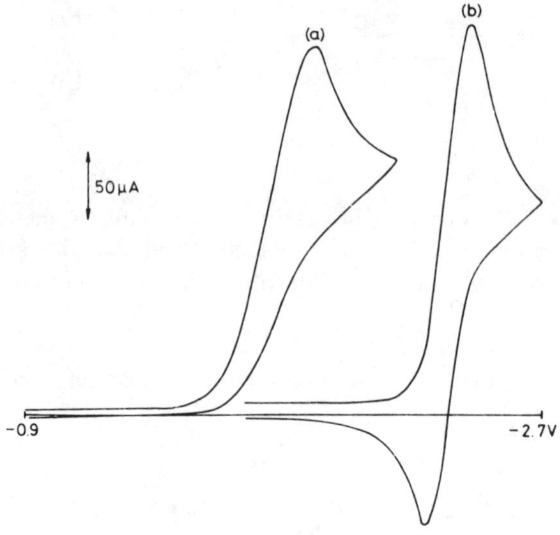

FIGURE 3.   Cyclic voltammograms of (a) PhClCN$_2$ and (b) Ph-(C$_4$H$_9$)CN$_2$ in MeCN-(C$_2$H$_5$)$_4$NClO$_4$ (0.1 M) at 50 mV/sec. (From Elson, C. M. and Liu, M. T. H., *J. Chem. Soc. Chem. Commun.*, 415, 1982. With permission.)

The reduction of the diaziridinone is again accompanied by cleavage of the N–N bond. Cyclic voltammetry was employed to determine if the initial reduction step proceeded via the radical (**5**):

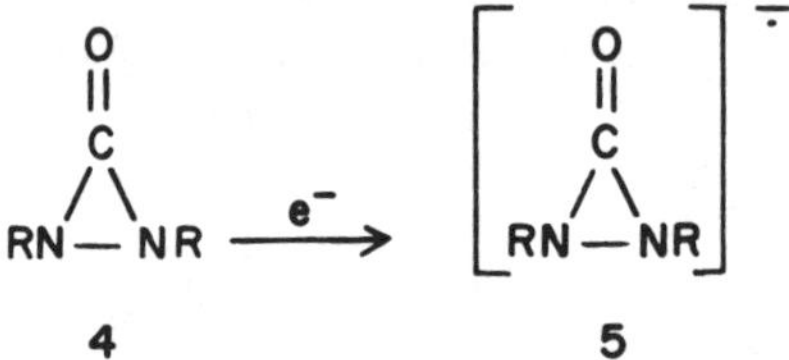

Using potential scan rates of up to 200 V/sec, (**5**) was not observed, i.e., an anodic wave was absent on the reverse scan, which implied that if (**5**) were formed it had a half-life of less than 2 msec.[14]

B. Recent Work

In the last few years, the electrochemical behavior of a series of 3-aryl-3-chlorodiazirines and 3-alkyl-3-phenyldiazirines in protic and aprotic media has been studied.[7,8] The cyclic voltammograms of representatives of these two types of diazirines appear in Figure 3. The voltammogram of PhClCN$_2$ is typical of the 3-aryl-3-chlorodiazirines and demonstrates an irreversible, diffusion controlled, one-electron reduction step. The peak potential, Ep$_c$, is -1.87 V (vs. Ag/AgNO$_3$ in MeCN) for the phenylchloro derivative. Similar behavior is observed for the (p-MeOC$_6$H$_4$)ClCN$_2$ compound except that Ep$_c$ shifts cathodically to -2.05 V. The shift reflects the difference in inductive effects of the electron donating p-methoxy group relative to hydrogen. Upon controlled potential electrolysis, both PhClCN$_2$ and (p-MeOC$_6$H$_4$)ClCN$_2$ are reduced by one electron per molecule to the corresponding arylnitriles, X-C$_6$H$_4$CN (X=H or p-MeO). The addition of 0.1% acetic acid to the electrolyte yielded the same electrode reactions and the same products.

Attempts to study (p-ClC$_6$H$_4$)ClCN$_2$ were unsuccessful, because only a transient reduction wave near -2.0 V was observed, i.e., the compound apparently reacts within minutes with the electrolyte system. Furthermore, (p-NO$_2$C$_4$H$_6$)ClCN$_2$ displays atypical behavior: two reduction waves are observed at -1.3 and -1.4 V and the second wave is reversible. It is known that the electron transfer pathway often changes when nitro groups are incorporated into molecules, and this was suspected in the present case. The reference compound, p-nitrochlorotoluene, exhibited a reduction wave at -1.9 V, indicating that reduction of the nitro group was probably occurring in both the diazirine as well as the substituted toluene.

By contrast, the 3-alkyl-3-phenyldiazirines (alkyl = n-C$_4$H$_9$ and CF$_3$) demonstrate the behavior depicted by curve b of Figure 3. At relatively fast potential scan rates (1 V/sec) or low temperature (0 to 5°C), the systems approach reversibility and i$_a$/i$_c$ approaches unity. As the temperature is increased or as the scan rate is slowed, the ratio of i$_a$/i$_c$ becomes smaller as the anodic wave height decreases. These observations are typical of a chemical reaction following the charge transfer reaction. The reduction step involves a single electron and is diffusion controlled; hence, the reduced species is characterized as the simple radical anion of the diazirines (**6**)

$$\text{R, R'}-\underset{\text{C}}{\big\langle}\,\begin{smallmatrix} N \\ \| \\ N \end{smallmatrix} \quad \xrightarrow[\text{MeCN}]{+e^-} \quad \left[\text{R, R'}-\underset{\text{C}}{\big\langle}\,\begin{smallmatrix} N \\ \| \\ N \end{smallmatrix}\right]^{\cdot -}$$

6

where R = C$_6$H$_5$ and R$'$ = n-C$_4$H$_9$ or CF$_3$. Such radicals which represent a new class of nitrogen containing radicals are strong oxidizing agents since they are formed at negative potentials (Ep$_c$ = -2.55 V, R$'$ = n-C$_4$H$_9$ and -1.97 V, R$'$ = CF$_3$). It is assumed that the observed chemical reaction involves the attack by the radicals on a component of the solvent supporting electrolyte. Assuming that the chemical reaction is pseudo-first order, the rate constants and halflives are 8 $\pm$ 1 $\times$ 10^{-2} and $\langle$2 $\times$ 10^{-2}/sec and 9 $\pm$ 1 and$\rangle$ 35 sec at 16°C for the n-butyl and trifluoromethyl compounds, respectively, as determined by the cyclic voltammetric and double potential step chronoamperometric methods. Such slow reactions explain why the voltammograms approach reversibility. The CF$_3$ group through its inductive effect shifts the reduction potential of the aryl-trifluoromethyl compound anodically by 0.6 V making it a weaker oxidizing agent than the n-butyl compound which may account for its greater half-life in solution.

The difference in the electrochemical behavior of the 3-aryl-3-chloro- and the 3-alkyl-3-phenyldiazirines is attributed to differences in the stabilities of the reduced species, since both types of diazirines are reduced over a comparable potential range. For the chlorodiazirines, the electrode reaction is irreversible because a chemical rearrangement coincides with the electron transfer. The rearrangement involves the loss of chloride and the opening of the CNN ring to produce the stable arylnitriles. Chloride is a good leaving group; by contrast, n-butyl and trifluoromethyl groups are poor leavers and, hence, when the alkylaryl compounds are reduced the structure of the compounds remains intact and the radical anions are formed.

Controlled potential electrolyses of the alkylphenyl diazirines in MeCN-(C$_2$H$_5$)$_4$NClO$_4$ at 5 to 8°C do not proceed smoothly to completion. The number of electrons transferred per molecule is noninteger and GC-MS analysis of the product solution of the electrolysis of (C$_6$H$_5$)(CF$_3$)CN$_2$ indicates that several compounds are present, including (C$_6$H$_5$)CO(CF$_3$) and (C$_6$H$_5$)CO(CH$_3$). This supports the earlier contention that the radicals react with the medium in the chemical step following the electrode reaction. Furthermore, during the early stages of the electrolyses, the initially colorless solutions of 3-n-butyl-3-phenyldiazirine (**6**)

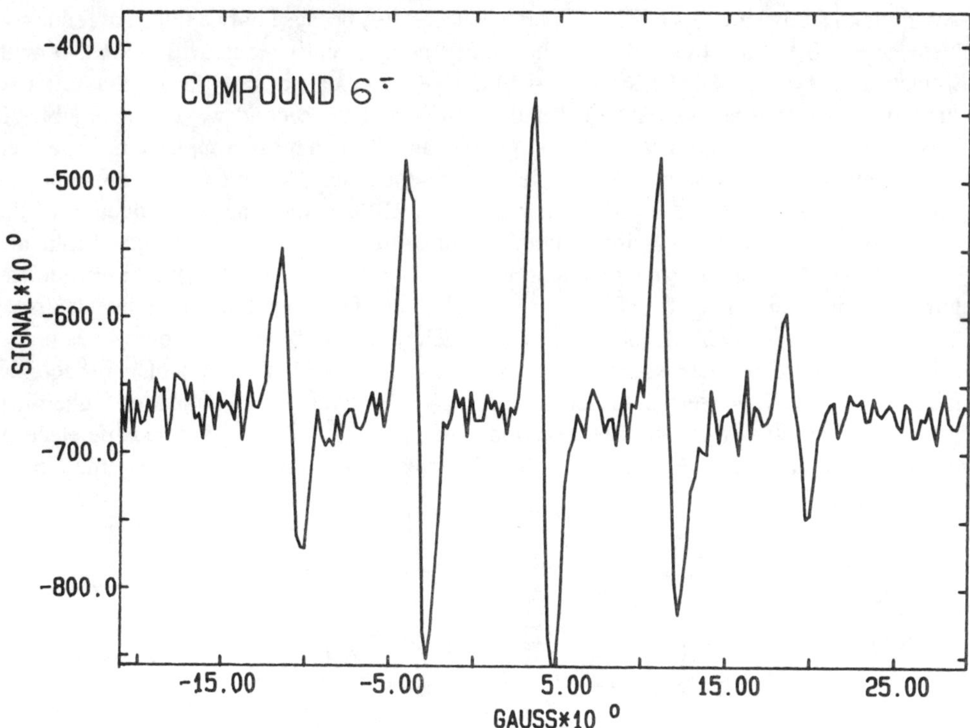

FIGURE 4.   ESR spectra of $[(C_6H_5)(C_4H_9)CN_2]^{\overline{\cdot}}$ (6⁻) at 10°C.

and 3-trifluoromethyl-3-phenyldiazirine (**7**) turned purple and yellow brown, respectively. It was assumed that these colored species were the anion radicals and since they remained in solution for several minutes it was possible to draw aliquots of the electrolysis solutions through glass tubing into the cavity of an ESR spectrometer and record their spectra. These spectra appear in Figures 4 and 5.

The spectrum of (**6⁻**) (the radical anion of the n-butylphenyl diazirine) is a simple quintet which arises from the coupling of the added electron to the two nitrogen atoms (S = 1), $a^N = 7.55 \pm 0.01$ g. The equivalence of the two nitrogens confirms the cyclic voltammetric findings that the diazirine ring is intact in the radical. Furthermore, the relatively large gauss value (2.0201) implies that a large portion of the unpaired spin density resides on the nitrogen atoms. The reaction of the radical with the medium is responsible for the deviation of the peak intensity pattern from the theoretical 1:2:3:2:1 pattern. In fact, the pseudo-first-order rate constant calculated from the decrease in the peak-to-peak height of the line at the highest field compared to the height of the lowest field line and the time of the spectrum scan was 2.0×10^{-2}/sec ($t_{1/2} = 35$ sec) at 10°. These values are consistent with the rate constant measured by the two electrochemical methods discussed earlier.

The spectrum of (**7⁻**) is a quintet of quartets. The quintet is again due to coupling of the added electron to the two equivalent nitrogens in the ring, and then each of these lines is split into four by the three equivalent fluorines (S = $^1/_2$). The hyperfine splitting constants are $a^N = 7.06 \pm 0.06$ gauss and $a^F = 1.76 \pm 0.06$ gauss; the smaller value of a^N for (**7⁻**) is due to the strong inductive effect of the CF_3 group compared to the n-butyl group of (**6⁻**). The peak intensity pattern closely approaches theoretical behavior. However, when the spectrum of a given sample is repeatedly recorded over a period of minutes, the disappearance of the signal is clearly marked. The rate of disappearance of the central, spectral lines can be used to determine the first-order rate constant for the reaction of (**7⁻**) with the supporting

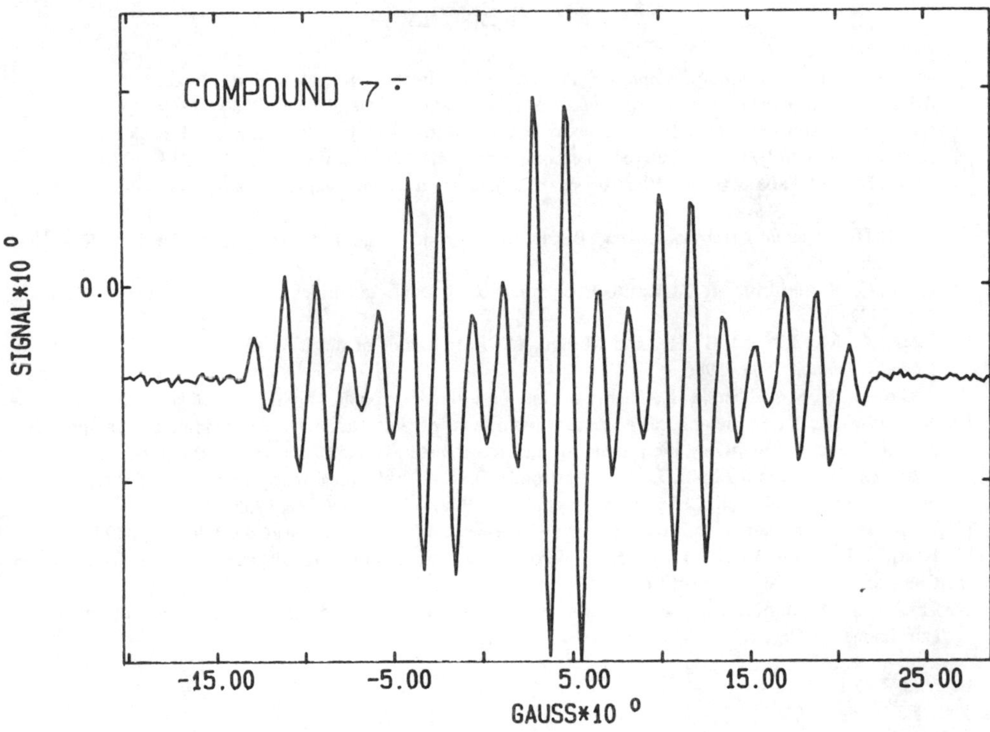

FIGURE 5.   ESR spectra of $[(C_6H_5)\,(CF_3)CN_2]^{\cdot}$ $(7^{\cdot})$ at 10°C.

electrolyte. The rate constant (1.5×10^{-2}/sec, $t_{1/2} = 46$ sec at 10°) is small, and explains why the cyclic voltammograms of (**7**) appear reversible. The most important result of the ESR work is the confirmation that the alkylphenyl diazirines undergo a reversible, single-electron reduction to form the simple anion radicals.

When a small amount of a proton source such as glacial acetic acid is added to the electrolyte, the electrochemical behavior of the alkylphenyl diazirines is altered significantly. Cyclic voltammograms appear irreversible at all potential scan rates, and the reduction peak potentials shift positively by 0.1 V. Moreover, the product of the controlled potential reduction of n-butylphenyldiazirine is the corresponding diaziridine, $(C_6H_5)(C_4H_9)CN_2H_2$. Hence, the presence of a source of H^+ alters the mechanism of the electrode reaction and the reduction of the nitrogen-nitrogen double bond is observed.

The electrochemical aspects of diazirine, a relatively new topic in the field of organic electrochemistry, offers many opportunities for research. The authors hope that this chapter will stimulate the exploration of the electrochemical behavior of these promising molecules.

REFERENCES

1. **Meites, L.,** *Polarographic Techniques,* 2nd ed., Wiley Interscience, New York, 1965.
2. **Adams, R. N.,** *Electrochemistry at Solid Electrodes,* Marcel Dekker, New York, 1969.
3. **Bard, A. J. and Faulkner, L. R.,** Electrochemical Methods, John Wiley & Sons, New York, 1980.
4. **Lund, H.,** Electrolysis of N-heterocyclic compounds, *Adv. Heterocyc. Chem.,* 12, 213, 1970.
5. **Lund, H. and Tabakovic, I.,** Electrolysis of N-heterocyclic compounds. II., *Adv. Heterocyc. Chem.,* 36, 235, 1984.
6. **Lund, H.,** *Organic Electrochemistry,* Baizer, M. M. and Lund, H., Eds., Marcel Dekker, New York, 1983, chap. 17.
7. **Elson, C. M. and Liu, M. T. H.,** Electrochemical behaviour of diazirines, *J. Chem. Soc. Chem. Commun.,* 415, 1982.
8. **Elson, C. M., Liu, M. T. H., and Mailer, C.,** ESR studies of diazirine anion radicals, *J. Chem. Soc. Chem. Commun.,* 504, 1986.
9. **Nicholson, R. S. and Shain, I.,** Theory of stationary electrode polarography, *Anal. Chem.,* 36, 706, 1964.
10. **Nicholson, R. S.,** Semiempirical procedure for measuring with stationary electrode polarography rates of chemical reactions involving the product of electron transfer, *Anal. Chem.,* 38, 1406, 1966.
11. **Schwarz, W. M. and Shain, I.,** Investigation of first-order chemical reactions following charge transfer by a step-functional controlled potential method, *J. Phys. Chem.,* 69, 30, 1965.
12. **Lund, H.,** Polarography and reduction of a diazirine, *Coll. Czech. Chem. Commun.,* 31, 4175, 1966.
13. **Kitajev, J. P. and Budnikov, G. K.,** Polarographic reduction of some diaziridines, *Coll. Czech. Chem. Commun.,* 30, 4178, 1965 (in Russian).
14. **Fry, A. J., Britton, W. E., and Wilson, R.,** Electrochemical and chemical reduction of di-tert-butyldiaziridinone, *J. Org. Chem.,* 38, 2620, 1973.

INDEX